T0207284

SpringerBriefs in History of Science and Technology

More information about this series at http://www.springer.com/series/10085

Roberto Lalli

Building the General Relativity and Gravitation Community During the Cold War

Springer

Roberto Lalli
Max Planck Institute for the History
of Science
Berlin
Germany

ISSN 2211-4564 ISSN 2211-4572 (electronic)
SpringerBriefs in History of Science and Technology
ISBN 978-3-319-54653-7 ISBN 978-3-319-54654-4 (eBook)
DOI 10.1007/978-3-319-54654-4

Library of Congress Control Number: 2017948200

Printed on acid-free paper

This Springer imprint is published by Springer Nature
The registered company is Springer International Publishing AG
The registered company address is: Gewerbestrasse 11, 6330 Cham, Switzerland

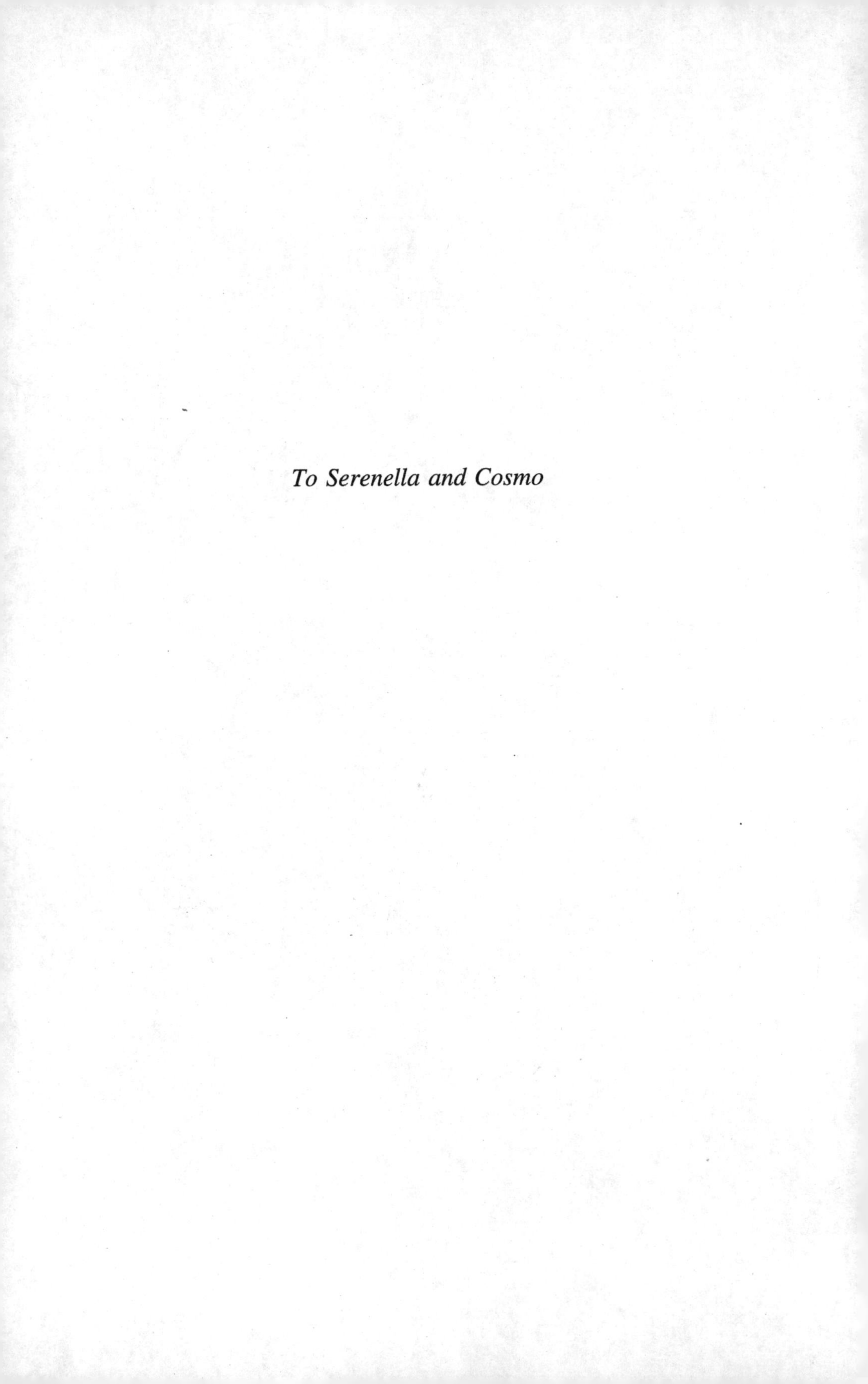

To Serenella and Cosmo

Acknowledgements

This book is one of the outcomes of a multi-institutional research project initiated and organized by Department I of the Max Planck Institute for the History of Science (MPIWG) aimed at exploring the historical process of the renaissance of general relativity. It could have never been written without the financial and organizational support provided by Department I of the MPIWG or the insightful discussion with the project participants. My greatest gratitude goes to the Director of the Department, Jürgen Renn, who made this research possible in many different ways ranging from institutional support to intellectual engagement.

Many of the arguments developed in this book and my historiographical approach have been elaborated in this collaborative environment, and the realization of this volume particularly owes much to the close cooperation with my colleague Alexander Blum and Jürgen Renn since the start of the project in 2014.

I am enormously grateful to a number of scholars who read an earlier version of the essay and made many insightful comments. I am especially indebted to the editor of the SpringerBriefs in the History of Science and Technology, Matteo Valleriani, who carefully followed the entire process from the presentation of a first incomplete version of this essay to its ultimate revision. The constant support of Dieter Hoffmann, also one of the editors of the SpringerBriefs in the History of Science and Technology, has been invaluable as he carefully read and commented on various versions of this book and substantially helped me in my research concerning the situation in East Germany and other Eastern Bloc countries. I also wish to give my heartfelt thanks to Michael Gordin, Jean-Philippe Martinez, and Benjamin Wilson whose comments helped me develop my arguments. I am very grateful to David Kaiser, as this book has been greatly influenced by the many discussions I had with him on the topic. The final version of this book attempts to respond to many of the comments, questions, and remarks these colleagues have made, and I hope that they might see the resulting improvements as their own contribution. All remaining errors, needless to say, are mine.

My heartfelt thanks go to all the participants of the research project "The Renaissance of General Relativity in the Post World-War-II Period" for the many illuminating discussions on the postwar development of general relativity as well as for having kindly provided some of the historical sources used in this research: Luisa Bonolis, Jean Eisenstaedt, Domenico Giulini, Hubert Goenner, Adele La Rana, Dennis Lehmkuhl, Christoph Lehner, Brian Pitts, Jim Ritter, David Rowe, Donald Salisbury, Tilman Sauer, Matthias Schemmel, and Jeroen van Dongen. The organizational support of Shadiye Leather-Barrow and Petra Schröter has been fundamental for the realization of the project. Additional material has been generously provided by Daniel Kennefick, and Dean Rickles, whom I thank very much.

Some arguments in this book were presented at the workshop "Space-Time Theories: Historical and Philosophical Contexts" at the Van Leer Jerusalem Institute in January 2015 and at the Centenary Conference on the History of General Relativity held at the Harnack-Haus in Berlin in December 2015. I am very grateful to all the organizers and attendees of both events for their comments and stimulating discussion, with special reference to Yemima Ben-Menachem, Diana Kormos-Buchwald, Hanoch Gutfreund, Carlo Rovelli, Robert Schulmann, and Aaron Wright. A very special thank you goes to Dirk Wintergrün who has been working on the technology for performing the network analysis of the general relativity and gravitation community, which inspired some of the arguments developed here. This book was completed during a two-month period spent at the Van Leer Jerusalem Institute, and I am enormously grateful to its previous and current directors, Gabriel Motzkin and Shai Lavi, for the financial and organizational support as well as for their interest in this project. Some of the arguments explored in this volume have been also inspired by the stimulating discussion within the Research Program on the History of the Max Planck Society of which I am a visiting scholar. I am very indebted to all the members of the Program for the illuminating views on the historical epistemology of scientific institutions developed within the program, including the Research Coordinator Florian Schmaltz and members of the Executive Committee Jürgen Kocka and Carsten Reinhardt, as well as Jürgen Renn.

To successfully pursue this research, I had the opportunity to access a number of historical documents stored in various archives and libraries around the world. This would have never been possible without the kind and extremely professional assistance of archivists, librarians, institute directors, and many dedicated experts I had the good fortune to meet. My sincerest thanks go to all the archivists and librarians who substantially helped me locate the materials, provided access to the documents, and gave permission to quote them. More specifically, I am immensely grateful to the following: Finn Aaserud, Helle Kiilerich, and Robert J. Sunderland (Niels Bohr Archive, Copenhagen) for Christian Møller's papers, institutional documents concerning the organization of the GR6 conference, and the related permission to use some excerpts and images in this book; Niklaus Bütikofer (Archives of the University of Bern) for some of the relevant documents concerning André Mercier; Gregory Good and Melanie Mueller (Niels Bohr Library & Archive, Center for the History of Physics, American Institute of Physics) for some documents belonging to the International Society on General Relativity and Gravitation; Diana Kormos-Buchwald and the staff of the Einstein Papers Project

(Pasadena, CA) for having provided access to some of the documents in the Einstein collection; Michael Miller (Archive of the American Philosophical Society, Philadelphia) for documents belonging to John A. Wheeler; Elisabeth Schlenk (library of the Albert Einstein Institute in Potsdam-Golm) for the papers of the International Society on General Relativity and Gravitation and for having organized the relocation of this material to the library of the MPIWG; Heffa Schucking for having lent the documents belonging to his father, Engelbert Schucking, to the library of the MPIWG; Natasha Swainston (Churchill Archives Centre, Cambridge) for having given access to the documents of Hermann Bondi, and for having provided permission to publish two images of this archival collection; the staff of the Syracuse University Archives (Syracuse University) for the permission to quote papers from the collection by Peter Bergmann. I wish to express my gratitude to Jean-Philippe Martinez for having discussed with me the content of the archival documents concerning Vladimir Fock's participation in the activities of the International Committee on General Relativity and Gravitation, preserved at the Archive of the Russian Academy of Science in St. Petersburg.

I wish to express my deep gratitude to Esther Chen, Urte Brauckman, Sabine Bertram, and all the staff in the library of the MPIWG for all the logistic support essential to the finalization of this book. The meticulous and dedicated work of the student assistants Felix Brümmer and Christopher Wasmuth has been of immense help throughout the entire research process. The final version of the text was produced with the invaluable editorial support of Linda Jayne Turner and Lindy Divarci. Words cannot express how indebted I am for all this priceless help in the production of this book. I am also very grateful to Lucy Fleet for the editorial assistance that allowed this essay to be published in the format of the SpingerBriefs.

My deep gratitude is for the persons to whom this book is dedicated, the "relativists" themselves, and their efforts to maintain historical records of this activity, starting with the former Secretary of the International Society of the General Relativity and Gravitation, Malcolm MacCallum, who has been extremely helpful in locating the society's historical materials. Georg Dautcourt, Joshua Goldberg, Kip Thorne, and Louis Witten have kindly shared important recollections on the events analyzed here and also provided insightful comments on the text itself. My thoughts go especially to those who made a major contribution to the evolution of the international General Relativity and Gravitation community and are no longer here to read this essay, particularly Vladimir Braginsky, Cécile DeWitt-Morette, and Felix Pirani, who sadly passed away while this book was being completed.

Certainly, every production is also the sum of the various experiences that implicitly or explicitly led to its realization. It is probably impossible to quote all the persons who have played an important part, in one way or another, in the very existence of this book. I wish only to mention my former mentor and Ph.D. advisor Pasquale Tucci, who introduced me to the history of physics and shaped my future career in many important ways, as well as my colleagues Massimiliano Badino, Leonardo Gariboldi, and John Stachel, who all had an impact on the evolution of my thoughts and research interests. Finally, my warmest thoughts go to my friends and my family who have always supported my studies in what is seen as a rather idiosyncratic discipline in Italy. I would like to thank my close friends Camilla Barbarito, Alberto Boccardi, Francesca Bonelli, Mila Casali, Giulia Damonte, Andrea Mei, my brother Emilio, and my sister Maria Irma. My deepest gratitude is for my late parents Irene and Manfredi, whose long-term, caring support I miss very much.

My last word of heartfelt thanks is for my wife Serenella who tolerated my mental absence and my poor sleeping while this book was under construction and she was pregnant with our first child. I see that this book is hers as much as it is mine.

Contents

Abbreviations

ASP	Alfred Schild Papers, 1915–1982, Briscoe Center for American History, University of Texas at Austin
APS	American Physical Society
BBAW	Berlin-Brandenburgische Akademie der Wissenschaften, Berlin
BOND	The Papers of Sir Hermann Bondi, GBR/0014/BOND, Churchill Archives Centre, Churchill College, Cambridge, UK
CDWP	Cécile DeWitt-Morette Papers, Briscoe Center for American History, University of Texas at Austin
CERN	European Council for Nuclear Research, Geneva
CMP	Christian Møller Papers, Correspondence 1971–81, Niels Bohr Archive, Copenhagen
CNRS	Centre National de la Recherche Scientifique, France
CPAE	Collected Papers of Albert Einstein, Einstein Papers Project, California Institute of Technology, Pasadena, CA
DAUT	Papers of George Dautcourt, personal collection, Berlin
DAWB	Deutsche Akademie der Wissenschaften zu Berlin (from 1972 Akademie der Wissenschaften der DDR)
DIAS	Dublin Institute for Advanced Studies, Dublin
ESP	Engelbert Schucking Papers, Library of the Max Planck Institute for the History of Science, Berlin
ETH	Eidgenössische Technische Hochschule, Zürich
GR6P	R6 Conference in Copenhagen, NORDITA Collection, Niels Bohr Archive, Copenhagen
HAM	Handakten Prof. André Mercier, 1934–1998, Staatsarchiv Bern, Bern
IAS	Institute for Advanced Studies, Princeton, NJ
IAU	International Astronomical Union
ICGRG	International Committee on General Relativity and Gravitation
ICSU	International Council of Scientific Unions
IGY	International Geophysical Year

IMU	International Mathematical Union
IOFP	Institute of Field Physics, University of North Caroline, Chapel Hill, NC
ISGRG	International Society on General Relativity and Gravitation
ISGRGR	International Society on General Relativity and Gravitation Records, 1961–1982, AR235, American Institute of Physics, Niels Bohr Library & Archives, College Park, MD
IUBS	International Union of Biological Sciences
IUGG	International Union of Geodesy and Geophysics
IUGS	International Union of Geological Sciences
IUPAC	International Union of Pure and Applied Chemistry
IUPAP	International Union of Pure and Applied Physics
JWP	John Archibald Wheeler Papers, 1880–2008, Mss.B.W564, American Philosophical Society, Philadelphia, PA
MATS	Military Air Transport Service
PBP	Peter Bergmann Papers (unprocessed collection), University Archives, Special Collections Research Center, Syracuse University Libraries, Syracuse, NY
PISGRG	Papers of the International Society on General Relativity and Gravitation (unprocessed collection), Library of the Max Planck Institute for the History of Science, Berlin
SGC	Soviet Gravity Committee
SPS	Swiss Physical Society
Technion	Israel Institute of Technology, Haifa, Israel
URSI	International Union of Radio Science
VFP	Vladimir Fock Papers, 1919–1974, Fond 1034, Inventaire 2, Archive of the Russian Academy of Sciences, St. Petersburg Branch, St. Petersburg
ZIAP	Zentralinstitut für Astrophysik, Deutsche Akademie der Wissenschaften zu Berlin

Chapter 1
Introduction

Abstract This chapter introduces the main arguments of the book in relation to the current historiographical debates on the origin and character of the process dubbed "renaissance of general relativity." It is argued that the return of Einstein's theory of gravitation to the mainstream of physics in the post-World War II period was as much epistemic as it was social, for it involved the formation of an international community of scholars that coalesced around a newly created research field called "General Relativity and Gravitation." These community-building activities led to an increasing degree of institutionalization that turned a dispersed set of scientists in the early 1950s into a well-defined organization—named the International Society on General Relativity and Gravitation—in the mid-1970s. The various steps in this institutionalization process are summarized. Although they might appear to be a straightforward incremental development corresponding to the establishment of a new scientific field and its subsequent growth, it is shown that all the various steps were extremely controversial. The book focuses on two elements that led to many of the tensions related to the international community-building activities: the uncertain epistemic status of general relativity at the time, and the developments of the Cold War, which deeply affected institutional processes in the international arena.

Keywords Albert Einstein · Cold War · Community building · Epistemic shift · General relativity · International Committee on General Relativity and Gravitation · International relations · International Society on General Relativity and Gravitation · Scientific institutions · Renaissance of general relativity

"To Relativists throughout the World": it was in this manner that the Swiss theoretical physicist and secretary of the International Committee on General Relativity and Gravitation (ICGRG), André Mercier, named the addressees of an official letter he sent to about 300 scientists scattered across six continents in early November 1972 (see Appendix B).[1] Attached to the letter, the recipients received the draft of the proposed statute for establishing the International Society on General Relativity

[1]André Mercier to Relativists throughout the World, November 1972, ISGRGR.

R. Lalli, *Building the General Relativity and Gravitation Community During the Cold War*, SpringerBriefs in History of Science and Technology,
DOI 10.1007/978-3-319-54654-4_1

and Gravitation (ISGRG). Mercier's package was meant to inform all those who were on Mercier's list of scientists active in research fields related to general relativity that the decision had been made to establish an international society to promote their common scientific interests. For the proponents of the society, the most important feature of the new institutional venture was its democratic character, as made explicit by Mercier's request for comments and suggestions about the draft statute. Every scholar on Mercier's list of "scientists throughout the World active in the field of Theories of Relativity and Gravitation"[2] could, in principle, convey his or her opinion on the institutional framework being developed.[3] One year and two months later, the ISGRG was formally established as the international institutional body devoted to general relativity and gravitation—a function that the ISGRG still has up to this day.[4]

The establishment of a scientific society aimed at supporting international exchanges and collaborations on research related to Einstein's gravitational theory can be considered one of the most striking outcomes of what theoretical physicist Clifford Will (1986 on pp. 3–18, 1989) coined the "renaissance of general relativity": the return of Einstein's theory of gravitation to the mainstream of physics after a thirty-year period of stagnation beginning in the mid-1920s.

The general context itself in which the construction of this international society occurred is, however, the subject of intense historical debate. While most experts, be they physicists with an interest in their history or historians of science, agree that a tremendous shift in the relevance of research in the field of gravitation occurred in the post-World War II period, no consensus has so far been reached on which were the most relevant causes sparking the renewal of research on relativistic theories of gravitation from the 1950s onward. Different authors identify different causes, indicating that each author has a different view of what this renaissance actually meant or was (see Chap. 2).

Recently, Alexander Blum, Jürgen Renn, and the author proposed a historical framework for interpreting this renaissance process as an outcome of the interplay between epistemic and social factors. From our perspective, the post-World War II transformations in the social dimension of theoretical physics and the newly created conditions for the transfer of knowledge played an important role both in launching the renewed interest in the field of gravitational theory and in shaping how this growing interest gradually evolved into a successful domain of research. Favored by changes in the environmental conditions, the return of general relativity to the

[2]Mercier to Scientists throughout the World active in the field of Theories of Relativity and Gravitation, January 1961, ISGRGR.

[3]Although a minority, at least ten female scholars were included in the list of those working in fields related to general relativity, the majority of whom were of French nationality. Some of them, like Marie-Antoinette Tonnelat, Cécile Morette-DeWitt, and Yvonne Bruhat, also had a prominent role within this community at the time.

[4]The first announcement the society had been established is in Mercier to Christian Møller, 7 January 1974, PISGRG, folder 1.4; the official announcement was in Mercier to Members of the ISGRG, 1 February 1974, PISGRG, folder 1.4.

mainstream of physics was characterized by a collectively shared recognition that what was required was an analysis of the still unexplored potential of the theory of *general relativity proper*. We refer here to the original theory of gravitation as formulated by Albert Einstein (with or without the cosmological constant), devoid of any attempt to consider it only as a stepping stone toward a more encompassing or fundamental theory (Blum et al. 2015, 2016).

As a process, this recognition was as much epistemic as social, for it entailed the formation of a community of scholars (including mainly physicists and mathematicians, as well as astronomers and, from the late 1950s, astrophysicists) that coalesced around a newly created research field. From the late 1950s, this field was named "General Relativity and Gravitation," or GRG for short. The self-organization of the community finally made it possible to collectively recognize the different research agendas related to relativistic theories of gravitation as part of an overarching field, which came to become more and more institutionalized at the international level.

The major steps of this institutionalization process are easy to identify. They began as early as November 1953, when plans were made to organize a conference in Bern to celebrate the fiftieth anniversary of the formulation of special relativity where Einstein had formulated the theory while working as a third class expert technician at the Swiss Federal Patent Office. Held in July 1955, the conference turned out to be the first ever international meeting entirely devoted to subjects related to general relativity (Mercier and Kervaire 1956). The Bern conference was the precursor to a long and stable tradition of international conferences dedicated to the newborn research area GRG.[5] It was followed by the Chapel Hill conference in North Carolina, USA, in 1957—where the community of American scholars working on general relativity and its quantization first played a major role—and the Royaumont conference, near Paris, in 1959 (DeWitt and Rickles 2011; Lichnerowicz and Tonnelat 1962). There, a group of scientists decided to establish the International Committee on General Relativity and Gravitation (ICGRG), whose main task was to organize large international conferences of this kind every three years. In one of the first meetings of the ICGRG, the members also agreed to establish a journal titled the *Bulletin on General Relativity and Gravitation* (*Bulletin on GRG*). This journal was not intended to contain scientific papers but, rather, it was envisaged only as a means of spreading useful information to be sent to those on Mercier's list of scientists "active in the field of Theories of Relativity and Gravitation."[6]

The *Bulletin on GRG* was published from 1962 to 1970, after which it was absorbed by a new scientific periodical called *General Relativity and Gravitation*

[5]For its role as the starting point of this long-lasting tradition, the Bern conference would later be known as "GR0," when the international conferences on the GRG field assumed the name of GR conferences. See "The GRn conferences," http://www.isgrg.org/pastconfs.php. Accessed 7 March 2016.

[6]Mercier to Scientists throughout the World active in the field of Theories of Relativity and Gravitation, January 1961, ISGRGR.

published under the auspices of the ICGRG, whose main task was to publish scientific papers in the field. *General Relativity and Gravitation* was the first scientific periodical specifically dedicated to publishing research in GRG (Mercier 1970). One year later, during the international conference held in Copenhagen, an ad hoc assembly of participants resolved to transform the institutional framework into an international society whose members would elect its board of directors.[7] After being formally established in January 1974, the ISGRG was almost immediately included within the larger structure of the International Union of Pure and Applied Physics (IUPAP).[8]

At first glance, this development might appear to be a straightforward institutionalization process: a progressive development from a set of dispersed research groups working on topics related to Einstein's theory of gravitation to a very structured international collective organization. This could be interpreted as a trivial consequence of the establishment of a new scientific field and its subsequent growth, both in terms of people involved and its relevance in the larger scientific community. As I shall show, however, the process was anything but straightforward. The actors themselves perceived many of these steps as extremely controversial for different reasons, both epistemic and political. The development of international relations during the Cold War in particular had a tremendous impact on how this institutionalization process unfolded. Eventually, political considerations came to determine the final structure of the International Society on General Relativity and Gravitation itself.

Historians of science David Kaiser and Benjamin Wilson (Wilson and Kaiser 2014; Kaiser 2015) have already persuasively argued that politics had an extremely relevant role in the historical evolution of a field such as general relativity—which is commonly perceived as one of the scientific domains closest to the ideal world of pure science. In the postwar period, science could not really remain aloof from the world of politics, as military funding and military-related research heavily influenced scientific production in virtually every field of scientific knowledge. Kaiser and Wilson's work is part of the recent attempts by historians of science to elucidate the way in which the *context* of the Cold War affected the *content* of the knowledge being produced in that context (see Oreskes 2014; and other essays in Oreskes and Krige 2014). The present book aims instead at bringing to light a different aspect of Cold War science: the complex impact that the developments of the Cold War had on how a specific group of scientists attempted to organize themselves in order to promote their field of interest. From the perspective of institutional history, it is not surprising that the developments of the Cold War affected the evolutions of the international bodies created for the purpose of promoting the field of general relativity and gravitation. What is relevant here is the specific ways in which the conflict between the political context and attempts by scientists to maintain their

[7]André Mercier to Relativists throughout the World, November 1972, ISGRGR.

[8]Mercier to Christian Møller, 7 January 1974, PISGRG, folder 1.4; and Anon. (1992, p. 37).

field "unpoliticized" evolved.[9] The establishment of the ISGRG in 1974 was the last step of the long journey to shape the institutional representation of the international community of "relativists." What in the mid-1950s was still a set of isolated scholars wishing to strengthen contacts to continue their research had become a well-identifiable community with a robust organizational structure by the mid-1970s. The established community had, however, radically changed how its members addressed political matters along the way. This book is the history of the dramatic route to the establishment of this community.

In reconstructing the history of the international institutional representations of the GRG community, this essay focuses on two elements. The first concerns the uncertain epistemic status of general relativity at the time, which made it problematic to consider the different research agendas entering the GRG field as belonging to a well-defined disciplinary domain. The second one is about how the developments of the Cold War deeply affected institutional processes in the international arena as well as the behavior and expectations of the scientists involved in the endeavor of community building. I argue that both these elements and the interplay between them had a strong impact on the forms that the international institutions under construction assumed.

The book is organized as follows. In Chap. 2, I will discuss the process of the renaissance of general relativity and its historiographical interpretation with a special focus on the integrated narrative put forward by Alexander Blum, Jürgen Renn and myself (Blum et al. 2015, 2016; see also Blum et al. 2017). In Chap. 3, I will introduce the broader context of how the re-establishment of international scientific exchange in the postwar and Cold War periods was generally structured. In the fourth and fifth chapters, I will provide a detailed historical narrative of the international institutionalization of the field of general relativity and gravitation against the backdrop of the transformations of international political relations from the mid-1950s to the mid-1970s. This historical evolution can be divided in two periods. The first one (Chap. 4), spanning from the mid-1950s to the mid-1960s, could be interpreted as the formative phase of the emerging community, during which initial steps were undertaken to institutionally unify the different research agendas under the heading of General Relativity and Gravitation. The second period (Chap. 5), from the mid-1960s to the mid-1970s, could be referred to as the maturity stage in which scientists attached to the existing institutional structure were involved in a variety of controversies. During these discussions, scientists endeavored to demarcate a clear boundary between scientific and political matters. In the attempt to do so, the participants came to hold very different views about where these boundaries should be drawn in the specific context of establishing an international scientific institution during the Cold War. Most actors involved in the construction and development of the ICGRG were very sensitive to the political situation and this

[9] For historical discussions about the intermingling of science and politics in the Cold War period see, for example, Wang (2008) and Wilson (2015).

attitude was beneficial, if not necessary, to the efforts to build an international community at that time. Other political factors had detrimental effects as they threatened the very existence of the recently established ICGRG. This period ended with the transformation from the ICGRG to the ISGRG—whose statute came to embody the political tensions characterizing its establishment. In the conclusion (Chap. 6), I will argue that the institutional history of this field and the final structure of the ISGRG was highly unusual, if not unique, in the institutionalization process of international scientific collaborations in the post-World War II period and I will offer possible explanations as to why the field of general relativity and gravitation favored the establishment of this rather original institutional structure.

References

Anon. 1992. *UIPPA-IUPAP 1922-1992*. Album souvenir realized in Quebec by the Secretariat of IUPAP. http://iupap.org/wp-content/uploads/2013/04/history.pdf. Accessed 7 March 2016.

Blum, Alexander, Roberto Lalli, and Jürgen Renn. 2015. The reinvention of general relativity: A historiographical framework for assessing one hundred years of curved space-time. *Isis* 106: 598–620.

Blum, Alexander, Roberto Lalli, and Jürgen Renn. 2016. The renaissance of general relativity: How and why it happened. *Annalen der Physik* 528: 344–349. doi:10.1002/andp.201600105 .

Blum, Alexander, Domenico Giulini, Roberto Lalli, and Jürgen Renn. 2017. Editorial introduction to the special issue "The Renaissance of Einstein's Theory of Gravitation". *The European Physical Journal H* 42: 95–105. doi:10.1140/epjh/e2017-80023-3.

DeWitt, Cécile M., and Dean Rickles (eds.). 2011. *The role of gravitation in physics: Report from the 1957 Chapel Hill Conference*. Berlin: Edition Open Access.

Kaiser, David. 2015. Cold War curvature: Measuring and modeling gravitational systems in postwar American physics. Talk presented at the conference *A Century of General Relativity*, Berlin, 4 December 2015.

Lichnerowicz, André, and Marie-Antoinette Tonnelat (eds.). 1962. *Les théories relativistes de la gravitation*. Paris: Éd. du Centre national de la recherche scientifique.

Mercier, André. 1970. Editorial. *General Relativity and Gravitation* 1: 1–7. doi:10.1007/BF00759197.

Mercier, André, and Michel Kervaire (eds.). 1956. *Fünfzig Jahre Relativitätstheorie, Verhandlungen. Cinquantenaire de la théorie de la relativité, Actes. Jubilee of relativity theory, Proceedings*. Helvetica Physica Acta. Supplementum IV. Basel: Birkhäuser.

Oreskes, Naomi. 2014. Introduction. In (Oreskes and Krige 2014), 1–9.

Oreskes, Naomi, and John Krige (eds.). 2014. *Science and technology in the global Cold War*. Cambridge, MA: The MIT Press.

Wang, Zuoyue. 2008. *In Sputnik's shadow: The President's Science Advisory Committee and Cold War America*. New Brunswick, NJ: Rutgers University Press.

Will, Clifford. 1986. *Was Einstein right?: Putting general relativity to the test*. New York: Basic Books.

Will, Clifford. 1989. The renaissance of general relativity. In *The new physics*, ed. Paul Davies, 7–33. Cambridge: Cambridge University Press.

Wilson, Benjamin. 2015. The consultants: Nonlinear optics and the social world of Cold War science. *Historical Studies in the Natural Sciences* 45: 758–804. doi:10.1525/hsns.2015.45.5.758.

Wilson, Benjamin, and David Kaiser. 2014. Calculating times: Radar, ballistic missiles, and Einstein's relativity. In (Oreskes and Krige 2014), 273–316.

Chapter 2
The Renaissance of General Relativity: A New Perspective

Abstract This chapter presents a general historiographical framework for interpreting the renaissance of general relativity as a consequence of the interplay between internal and environmental factors. The internal factors refer to the resilient theoretical framework provided by general relativity to physicists working in diverse and dispersed fields. The external factors relate to the changing working conditions of physicists in the post-World War II period, with the newly created conditions for the mobility of young researchers, for the transfer of knowledge in a growing international community, and for the self-organization of an identifiable community. These external factors created a favorable environment for integrating the dispersed research endeavors under the new heading of "General Relativity and Gravitation" research. This, in turn, provided the conditions for the emergence of a coherent investigation of the theoretical core of general relativity for its own sake and for the creation of a community specifically dedicated to this goal.

Keywords Albert Einstein · Epistemic dispersion · General relativity · Low-water mark of general relativity · Quantization of gravity · Relativistic cosmology · Renaissance of general relativity · Unified field theory · Untapped potential of general relativity

What was the renaissance of general relativity? What were the main features of this phenomenon? What were its main phases? On what empirical foundations can we base our claim that there was a revitalization of the field of general relativity in the post-World War II period? Does the term "renaissance" really capture the many facets of this complex historical process?

These and other related questions are at the center of a lively debate among historians of modern science and physicists. The origin of this debate is to be found in the works of historian of physics Jean Eisenstaedt and physicist Clifford Will who identified two consecutive, and symmetrical, epochs in the history of general relativity. In his influential papers, Eisenstaedt maintained that the initial burst of

This chapter is based on the historiographical framework developed in Blum et al. (2015).

R. Lalli, *Building the General Relativity and Gravitation Community During the Cold War*, SpringerBriefs in History of Science and Technology,
DOI 10.1007/978-3-319-54654-4_2

excitement about the theory following the acclaimed 1919 announcement that one of its few empirical predictions—gravitational light bending near massive bodies—had been confirmed by authoritative British astronomers was short-lived. According to Eisenstaedt (1986, 1989), as of the mid-1920s, research about the theory of general relativity underwent a thirty-year period of stagnation. This situation, which Eisenstaedt called the "low-water mark" of general relativity, ended around the mid-1950s, when work on the theory began producing novel results at a higher pace and attracting a host of new research scholars.

In his reviews of the experimental tests of Einstein's theory of gravitation, physicist Clifford Will (1986, 1989) stressed that this activity did not emerge in full force until the late 1950s. In his view, by 1970, general relativity has become "one of the most active and exciting branches of physics" (Will 1989, p. 7)—a process that deserved the splendid title of "renaissance." It is striking that both these analyses were published in the second half of the 1980s, suggesting that by that time general relativity had attained the status of one of the building blocks of modern physics, together with quantum mechanics and quantum field theory. Practitioners probably felt that what was seen as a solid column in the edifice of physical knowledge in the mid-1980s stood on very shaky ground less than three decades previously. Following these early attempts to frame the post-WWII history of the theory and recent explorations of its early phase, a periodization of the history of general relativity was proposed (Gutfreund and Renn 2017):

1907–1915: The genesis of general relativity: this phase represents Einstein's search for a relativistic gravitational theory culminating in the final formulation of the equation of general relativity published on 25 November 2015 (Renn 2007).

1915–ca. 1925: Formative period of general relativity: this decade is marked by attempts to test the theory, its extension of application to cosmology, and early lively discussions within the physics and mathematics communities (Gutfreund and Renn 2017).

ca. 1925–ca. 1955: Low-water-mark period.

From ca. 1955: Renaissance of general relativity.

Although the dates are still a matter of debate, most experts usually agree that this view is a respectful representation of the historical trajectory of general relativity. We still need, however, to better identify the transition between the low-water-mark and the renaissance phases and define these two periods together, for neither the term "low-water mark," nor "renaissance" makes complete sense without a clear definition of the other. For Eisenstaedt, the main features of the low-water-mark phase were the following. Firstly, only a few scientists—mostly mathematicians—worked on the theory during this period. From the early 1920s, physicists lost interest in a theory that, on the one hand, was very complex from the mathematical standpoint and, on the other, had very little, if any, connection with experimental or observational research. It is not particularly surprising that most theorists preferred to focus on quantum mechanics and its plethora of applications to microphysics and solid-state physics. Contrary to general relativity, theoretical problems of quantum mechanics had direct and productive links to experimental activities. Besides the technical difficulties, research into general relativity was unattractive, for the general impression was that working on it would have led only to purely formal improvements or minor corrections to Newtonian physics.

From the conceptual standpoint, this state of affairs created a barrier to gaining a deeper understanding of the physical predictions of the theory. The meaning and physical characteristics of the Schwarzschild solution and the implications of general covariance for the notions of space and time, for instance, remained clouded with confusion up until the renaissance phase. Eisenstaedt maintains that the few who worked on the theory employed what he called the "neo-Newtonian interpretation" of the theory, particularly of the space-time coordinates. This implied that, during this period, scientists found it very difficult to draw a clear demarcation between actual predictions of the theory and artifacts due to the coordinates used, which were often chosen only to simplify calculations for specific problems. Implicitly, Clifford Will agrees with this view by stating that the theory moved away from being perceived as a highly formalistic subject to being considered one of the most exciting branches of physics by the late 1960s.

After the groundbreaking analyses by Eisenstaedt and Will, historians of science and physicists who discussed the post-World War II history of general relativity tended to agree with this general picture: an important shift in the relevance of research in general relativity occurred sometime around 1960 (see, e.g., Thorne 1994; Kragh 1999; Kaiser 2000; Kennefick 2007). These scholars either accept the term "renaissance" or use a similar characterization of the period, such as the "Golden Age" of general relativity (Thorne 1994).

With at least one important exception (Goenner 2017), there is strong agreement among scholars that a process that could be called "renaissance of general relativity" actually occurred after World War II. The same scholars, however, have provided quite different views about what the causes of this phenomenon were, which is related to different definitions of the renaissance, its periodization, and its main features.

2.1 Review of the Historiographical Debate

One explanation that has been proposed by Will and is usually highly respected by working physicists is that the revitalization of the interest in general relativity and its sudden progress was sparked by empirical confirmations and experimental discoveries, particularly in the astrophysical domain. The major events normally credited as having resurrected the field were the discovery of quasars in 1963, of the cosmic microwave background radiation (CMB) in 1965, and of pulsars in 1967. All these discoveries required a theory of gravitation that allowed these empirical phenomena to be analyzed and understood in a coherent theoretical framework. The theory of general relativity provided this framework as the currently accepted theory of gravitation. All the abovementioned discoveries were serendipitous and resulted from the rapid innovation in instrumentation. Therefore, the focus on the experimental impulse of the renaissance of general relativity implies that the

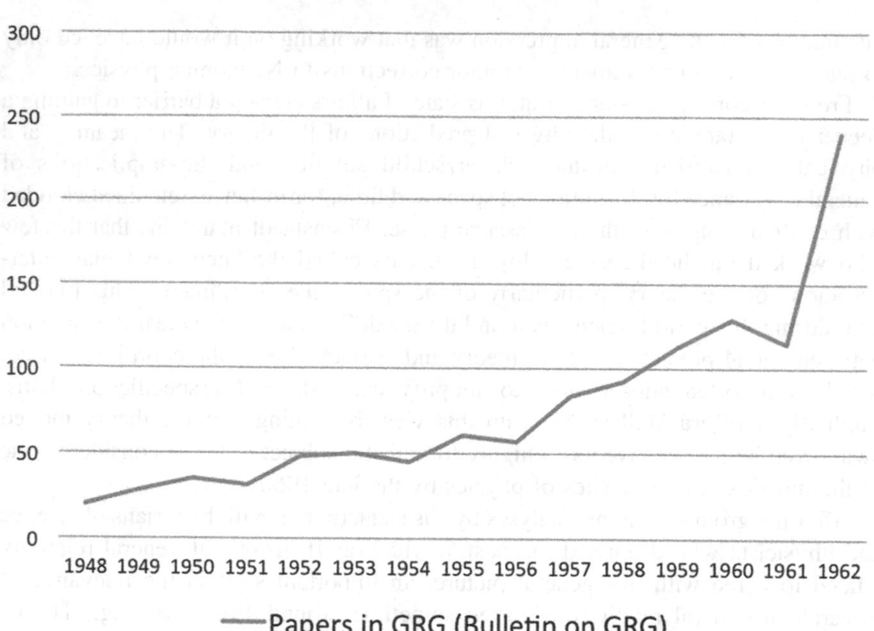

Fig. 2.1 Statistical analysis conducted by the author on the papers in the field of GRG published between 1948 and 1962 as listed in the *Bulletin on GRG*

process was a direct consequence of tremendous technological advances, mostly related to scientific research during World War II and the ensuing Cold War. Only thanks to these new technologies, so the argument goes, could the theory eventually find novel and successful connections with the empirical domain.[1]

Without doubt, these discoveries played a major role in shaping how theoretical gravitation research unfolded. Yet it does not explain other features of the process, such as, for instance, the extraordinary increase in the number of papers addressing topics related to Einstein's theory of gravitation during the 1950s. In Figs. 2.1 and 2.2, different criteria and methodologies have been applied to examine the changes in the number of scientific publications on subjects related to general relativity over the years.[2] Although the two diagrams do not agree on the details, they both strongly support the view that a substantial increment of the scientific production occurred *before* 1962, namely, one year before the discovery of quasars. This increment was therefore completely unrelated to the serendipitous astrophysical and astronomical discoveries cited above.

[1]For a review of these discoveries and their consequences, see Longair (2006). See also Peebles (2017) for a thorough discussion on the evolution of the experimental work in the field of gravitation from the late 1950s to the late 1960s—a period that Peebles calls the "naissance" of experimental gravity physics.

[2]See also the study published in Eisenstaedt (2006. p. 248).

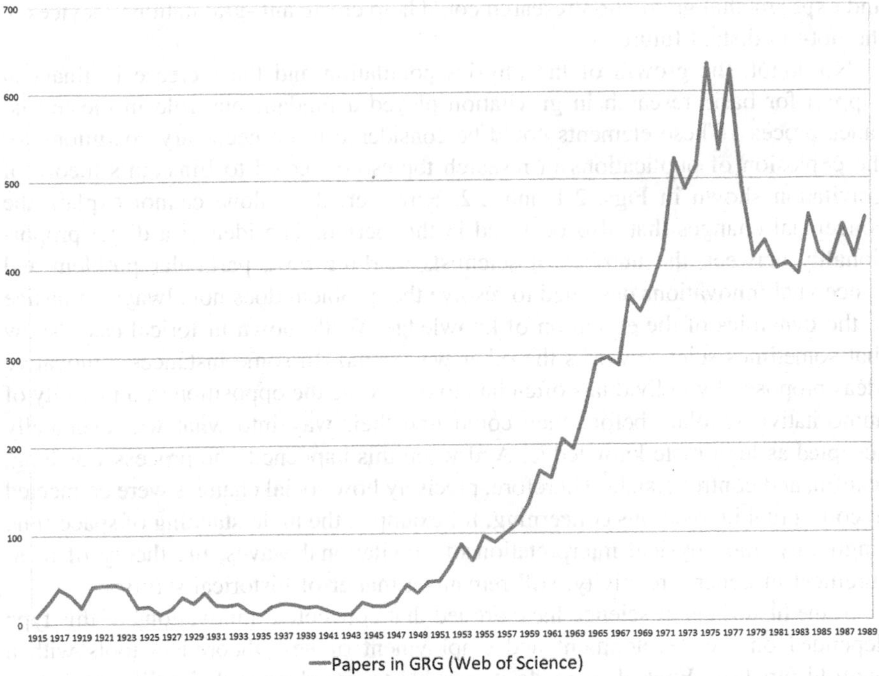

Fig. 2.2 Statistical analysis conducted by the author on papers on topics related to gravitation found in *Web of Science*

One explanation in agreement with the diagrams is that what Will called the "renaissance of general relativity" was a simple consequence of the enormous increase in the physics population after World War II, including of course the growth in the number of theoretical physicists.[3] In other words, this argument states that the proportion of the scientific community working in the field of general relativity did not change significantly after World War II. It was the abrupt change in the total number of active physicists that made a huge difference because this change implied much more far-reaching consequences in a field as small as general relativity was at the time.

An argument strongly related to the previous one is that this research field also benefited from the unprecedented flow of money going into basic science research in the post-World War II period, much of which came from military sources, especially in the United States. Recent studies by Kaiser (2000) and Rickles (2015) have convincingly shown that military funds and private patronage allowed the emergence and flourishing of research centers devoted to general relativity and related fields. The motivation for this generous support was often the hope that theoretical

[3]A very meticulous study of this demographic transformation in the United States was conducted by Kaiser (2012).

and experimental gravitation research could help create anti-gravitational devices in the not-too-distant future.

No doubt, the growth of the physics population and the increase in financial support for basic research in gravitation played a fundamental role in the renaissance process. These elements could be considered to be necessary conditions for the explosion of publications in research topics connected to Einstein's theory of gravitation shown in Figs. 2.1 and 2.2. However, they alone cannot explain the conceptual changes that also occurred in this period. The idea of a direct proportionality between the number of scientists working on a particular problem and conceptual innovations designed to resolve that problem does not always do justice to the dynamics of the evolution of knowledge. Well-known historical cases show that sometimes science works the other way round. In some instances, innovative ideas proposed by individuals often had to overcome the opposition of a majority of authoritative scholars before they could find their way into what was eventually accepted as legitimate knowledge. And when this happened, the process was long, painful, and controversial.[4] Therefore, precisely how social changes were connected to conceptual innovations concerning, for example, the understanding of space-time singularity, the physical interpretation of gravitational waves, the theory of measurement in general relativity, still remains a matter of historical scrutiny.

Some historians of science have argued that conceptual innovations of this type depended on the development and employment of new theoretical tools with a twofold function. First, they made it possible to speed up and simplify the calculations in general relativity, which, everyone agreed, were painstaking.[5] Second, some of these new tools led to improved visualization of the general relativistic space-time, allowing a clearer intuitive, physical interpretation of the theory. Many are the theoretical tools introduced between the mid-1950s and the early 1960s that, according to commentators and practitioners, were crucial to the most important conceptual advances related to the belief in entities such as gravitational waves and black holes. The most quoted are: (a) the Petrov classification (also called the Pirani-Petrov or Penrose-Pirani-Petrov classification) of the Weyl tensor, first published in 1954 (Petrov 2000); (b) the tetrad and spinor formulation developed mainly by the British mathematician Roger Penrose in the early 1960s (Penrose 1960; Newman and Penrose 1962); (c) the Kruskal-Szekeres coordinates elaborated by 1960 (Kruskal 1960; Szekeres 1960); and (d) the Penrose diagrams, which became a diffused tool during the 1960s. The only in-depth historical study on the development and relevance of these theoretical tools to date has been by historian of science Aaron Wright (2014) who argued that the Penrose diagrams did for general

[4]This kind of process might be considered a fundamental part of the concept of scientific revolution as defined in Kuhn (1970).

[5]Even the mathematically minded physicist Pascual Jordan complained about the "mismatch between the simplicity of the physical and epistemological foundations and the annoying complexity of the corresponding thicket of formulae" (Jordan 1955, p. 5, translated in Blum et al. 2017, p. 96).

relativity what the Feynman diagrams did for quantum electrodynamics.[6] More specifically, Wright stresses that the Penrose diagrams made it intuitively possible to grasp the meaning of space-time infinity in the theory of general relativity.

The view that new theoretical tools, particularly those quoted above, played a predominant role in the epistemic shift between the previous neo-Newtonian interpretation of the theory and a fully relativistic understanding of the general relativity theory is certainly consistent. Recollections of the protagonists also stress the relevance of one or more of these tools in allowing a deeper grasp of the extreme physical implication of Einstein's theory.[7] But no study to date has produced a detailed analysis of the conceptual changes in general relativity in terms of these theoretical tools. Moreover, the recourse to theoretical tools to explain the renaissance appears somehow tautological, for it does not explain the phenomenon itself of the emergence and use of these theoretical tools. It does not explain why these tools were all formulated in the period between mid-1950s and early 1960s and why they were soon successfully used by many scientists to produce conceptual advances which might be considered to be of a somewhat revolutionary nature as far as the predictions of actual physical phenomena is concerned.

Historical studies have in fact revealed that precursors of many of these tools were already available in the low-water-mark period. The work of French mathematician Cartan (1922a, b) on the classification of the Weyl spaces that led to the Petrov classification was largely ignored for decades. The work of American cosmologist and mathematical physicist Howard P. Robertson and others on the Schwarzschild singularity could have led to a better understanding of what we now call black holes much earlier than 1960 (Eisenstaedt 1987). The tetrad formalism was actually developed in the context of research in unified field theory in the 1920s (Goenner 2004). None of these precursory advances led to the same definition of problems in physical terms within general relativity proper as occurred between the mid-1950s and the early 1960s. Why this sudden increase in new theoretical tools explicitly created to deal with the problems of the theory of general relativity happened in this specific period remains unexplained.

2.2 Re-assessing the Low-Water-Mark Period

As I have shown in the previous sections, the different views proposed so far by historians of science and physicists leave some major questions unanswered, the main problem being that we still do not have a unified framework to describe the

[6]On the dissemination of the Feynman diagrams and their role in the evolution of theoretical physics, see Kaiser (2005).

[7]Ezra Newman and Roger Penrose, 13 December 2013, interview with Alexander Blum, Jürgen Renn, and Donald Salisbury; and Dieter Brill and Charles Misner, 13 December 2013, interview with Alexander Blum, and Donald Salisbury. I am very grateful to Alexander Blum, Jürgen Renn, and Donald Salisbury for having provided the records of these interviews.

renaissance process. An attempt to frame this unified narrative which takes into account both epistemic and social factors was recently made by Alexander Blum, Jürgen Renn and myself (Blum et al. 2015; see also Blum et al. 2016, 2017). To understand and define the period of the renaissance, the first step was to revise the concept of low-water-mark period. The low-water mark and the renaissance are in fact symmetrical historical categories and it is not possible to understand the renaissance without an in-depth discussion of what happened before.

In our view, one of the most striking features of the low-water-mark period—one that distinguishes it from the renaissance—has not been taken into consideration by previous historical analyses. If we look at the low-water-mark phase without taking the survival of the theory for granted, we see that those who worked on the theory pursued the main goal to modify general relativity and to replace it with a more encompassing one. They mostly aimed at formulating a theory able to describe different physical forces under the same theoretical framework. These manifold attempts were directed in particular toward the search for a unified field theory of the gravitational and electromagnetic phenomena (Goldstein and Ritter 2003; Goenner 2004, 2014). Einstein himself dedicated many years of research to this attempt (van Dongen 2010). Whereas, during the low-water-mark period, there were many approaches to this problem, the most diffused followed the methodology allegedly pursued by Einstein himself in his successful path toward the theory of general relativity. The geometrization of physics was perceived by many as the high road that could have led to a unified theory of gravitation and electromagnetism.

A second, minor, theoretical approach saw the gravitational field only as another field to be quantized following the success of quantum mechanics. These attempts first began in the early 1930s and produced a set of formal steps forward, but without any physical predictions (Blum and Rickles 2017). Both the programs on unified field theory and on the quantization of Einstein's equations made use of some principles of Einstein's gravitational theory as well as his heuristics and methodology. However, this was done with the goal of finding a superior theory, through attempts that were ultimately unsuccessful. The superior goal of going beyond Einstein's theory shaped the way scientists looked at general relativity and at its physical predictions. Those who worked on the above-mentioned research agenda did not consider Einstein's theory fundamental enough to warrant detailed scrutiny of its implications, nor did they think that the theory contained much empirical potential besides what was already known.

The major exception to this attitude was in cosmology. Between 1927 and 1933 there were numerous advances in the field of physical relativistic cosmology, which led to the formulation of the expanding universe. Research in this area was so advanced that in 1933 Howard P. Robertson published a review on relativistic cosmology presenting a basic model of the evolving universe, which is still considered part of the standard present-day Big Bang cosmological model (Ellis 2012, p. 2108). Even in the case of cosmology, however, these developments were received with skepticism by the majority of physicists who questioned whether cosmology was a scientific field at all. It did not help matters that controversies between founders of relativistic cosmology and proponents of alternative theories focused on somewhat philosophical and meta-scientific arguments concerning what was the most suitable method to make progress in a field so far from the observational domain as that of cosmology.[8] Consequently, the extreme physical implications of the theory, such as the primeval atom proposed by Lemaître in 1931, were distrusted by the majority of practitioners (Kragh and Lambert 2007).

During the low-water-mark period, mathematical advances in the area of gravitational theory continued to be pursued both within the program on unified field theory and as an independent research field in mathematics. An important result regarding the physical application of the theory was also obtained by Oppenheimer and his co-authors in their study of the gravitation of a collapsing star in 1939 (Bonolis 2017). All these advances did not become, however, a pool of knowledge shared by practitioners in the field. Most of the results that were considered of value with hindsight were at the time often ignored or distrusted (Ortega-Rodríguez et al. 2017). All these research agendas connected to Einstein's theory of gravitation, in fact, appear as a set of different approaches directed toward quite different goals where the only connection was that knowledge of general relativity and of specific mathematical tools was necessary in order to make progress. These activities were therefore characterized by a strong degree of epistemic dispersion, where scholars did not agree either on the goal or on the methodology. There was no common way to evaluate results, nor was it clear which discipline these results belonged to, whether it was pure mathematics, physics, astronomy, or astrophysics.

This kind of epistemic dispersion was accompanied by a strong social dispersion. Historical studies have revealed that a number of insights were gained in some research branches related to general relativity, particularly in the fields of cosmology and unified field theory (see, e.g., Goenner 2004, 2014; Eisenstaedt 2006), but this progress remained unrecognized in a strongly dispersed network of practitioners that was divided by disciplinary and national boundaries. The means of communication employed by scientists working on problems related to general relativity did not favor a smooth and rapid transmission of knowledge. Papers on these matters could be found in highly diverse publication venues in disciplines such as mathematics, astrophysics, astronomy, and physics. No conference specifically dedicated to exploring all aspects of general relativity, less alone a specific one, was ever organized before 1955.[9] In brief, it was not possible to identify a coherent community of practitioners with shared methods, research questions, and a similar language.[10] The

[8]See George Gale, "Cosmology: Methodological Debates in the 1930s and 1940s," *The Stanford Encyclopedia of Philosophy* (Spring 2014 Edition), ed. Edward N. Zalta, http://plato.stanford.edu/archives/spr2014/entries/cosmology-30s/. Accessed 21 September 2016.

[9]There were a few exceptions, however. Following some developments, the program of unified field theory was revitalized in the period 1929 to 1930 and unified theory also became one of the main topics at the first Soviet All-Union Conference on Theoretical Physics in Kharkov, Ukraine (Goldstein and Ritter 2003). Shortly afterwards, the program seemed be peripheral again (see Vizgin and Gorelik 1987, p. 312).

[10]I am referring in particular to the definition of a scientific field from the perspective of a collaboration network (see, e.g., Bettencourt et al. 2008).

dispersion of the activities in the macro-area broadly connected to general relativity, which, as we have seen, was as much epistemic as social, implies that no scientific field known as general relativity existed at all during the low-water-mark phase. There was no identifiable area of research to which practitioners could refer, or belong, in their pursuit of various research agendas.

2.3 Exploiting the Untapped Potential of General Relativity

These different, dispersed research traditions constituted a potential which was activated during the more favorable societal conditions of the post-WWII period. The abovementioned research activities were the foundations of what is broadly considered to be the successful return of general relativity to the mainstream of physics. The question that then arises is precisely how the existing potential for further developments was activated.

Albeit dispersed, the research traditions previously discussed kept interest in general relativity alive and acted as a conduit for the transmission of Einstein's theory to the next generation through research projects in fact aimed at going beyond Einstein's theory. This process led to a cascade of transformations of general relativity in the 1950s. A few research centers devoted to one or more of the various traditions going beyond general relativity were established around the mid-1950s. By research center, I refer to any kind of institution (universities, private or public research institutes, sections of scientific academies, etc.) where there was at least one principal investigator who had an institutional position stable enough to attract postdocs and/or produce new Ph.D.'s in the field (see Fig. 2.3 and

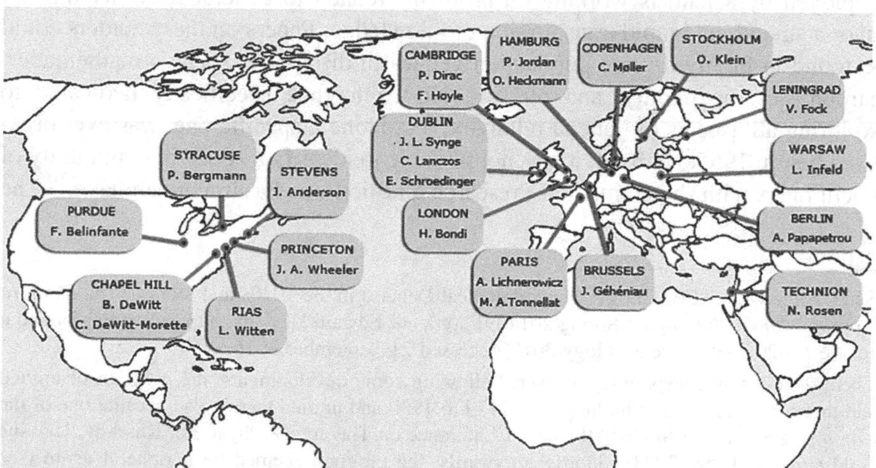

Fig. 2.3 Map of the major research centers working on topics related to general relativity in the United States and Europe in 1955 (see also Appendix A)

Appendix A for a list of the research centers working on research agendas connected to general relativity in 1955.) These research centers benefited enormously from the general transformation in the social dimension of physics that occurred in the post-World War II period, namely, the substantial increase in talent and money flowing into physics in general, and theoretical physics in particular.

Besides this general transformation, one element that seemed to play a specifically important role was the establishment and consolidation of the tradition of postdoctoral education. Given the demographic explosion of physics in the 1950s, many of the new Ph.D.'s did not, and could not, find a stable position in the academe immediately after graduating. They had to spend two or three years, or in some cases even longer, doing postdoctoral research in various research centers in more than one country. In fields related to general relativity, with almost nonexistent connections with industrial and military applications, this phenomenon was even more marked than in other branches of theoretical physics. The long pilgrimage of young researchers made it possible to establish links between the different research centers. The transfer of persons, in turn, facilitated the transmission of theoretical tools, concepts, and research questions between the different research centers and then from one research tradition to another. This process turned the dispersion of the activities into an asset as the developments pursued in different centers soon bore fruit in different contexts.[11] As we see it, this process was a major component in the reconstruction of knowledge giving rise to conceptual transformation in the field of general relativity.

This was not sufficient, however. A relevant role was also played by the explicit attempts to build a community of scientists working on what could be identified as the larger research domain, which included the various research agendas previously described. These attempts began around the mid-1950s and led, through a series of steps summarized in the introduction to this book, to the institutional establishment of General Relativity and Gravitation as a scientific field (see Chap. 1).

The new possibility of social interactions led the leaders of many research centers to identify and formulate common questions, which started to focus more and more on general relativity proper. Initially, this recognition occurred under the assumptions that it was necessary to explore the original theory in detail before furthering the different research programs, which aimed at modifying the theory or going beyond it. This conscious recognition was at the basis of a conceptual reconfiguration—an epistemic shift—where new shared questions concerning, for instance, the theory of observables in general relativity and the properties of gravitational waves became central to the various research agendas (Blum et al. 2017). And because of the social evolution toward a more structured community of scholars, this shift rapidly became a common feature of the newborn community at

[11]Kaiser (2005) studied this process in the context of the diffusion of the Feynman diagrams and called it the "postdoc cascade."

large.[12] This commonly shared change in the focus of research programs toward more conservative goals concerning Einstein's theory of gravitation in its own right was, in our view, the central mechanism of the renaissance of general relativity, which anticipated the new discoveries in astrophysics in the 1960s.

The role of these discoveries was then to bolster a process already well established. It is remarkable, in fact, that only nine months after the first announcement of the discovery of a new astrophysical object, soon to be named quasar, a large and successful conference, the First Texas Symposium on Relativistic Astrophysics, was held where the connections between this discovery and the possible explanation within the context of general relativity were explicitly drawn (Robinson et al. 1965). Even if general relativity was not immediately used to give a realistic physical description of the dynamics involved in the newly discovered astrophysical objects, it was accepted that it would be able to do so in future. In other words, it was acknowledged that the general physical mechanisms Einstein's theory proposed to describe, such as the formation of quasars, were correct, and consensus was rapidly built around such a belief. The speed with which this process occurred would have probably been inconceivable without a community of relativists prepared to absorb this discovery both epistemically and sociologically by organizing a large conference within a few months.

To summarize, in our recent work (Blum et al. 2015, 2016), we claimed that the phenomenon of the renaissance can be seen as a consequence of the interplay of what we categorize as internal and environmental factors. The internal factors refer to the resilient theoretical framework provided by general relativity to physicists working in diverse (and dispersed) fields; the external factors relate to the changing working conditions for physicists in the post-World War II period. Here, we do not only mean the availability of new technologies, the growing number of practitioners, and the exponential increase in funding. We also refer to newly created conditions for the mobility of young researchers (post-doc cascade), for the transfer of knowledge in a growing international community, and for the self-organization of an identifiable community. These external factors created a favorable environment for integrating the dispersed research endeavors under the new heading of GRG research. This, in turn, created the conditions for the emergence of a coherent investigation of the theoretical core of general relativity for its own sake and for the creation of a community specifically dedicated to this goal. This is also the sense in which Blum, Renn and I propose to speak not of the mere renewal of relativity research but of the reinvention of general relativity within the physics discipline, which was thus turned from a theoretical framework into a field of research in its own right.

Within the historical framework outlined above, this book explores a fundamental aspect of the social dimension of this process by focusing on the explicit

[12]This is confirmed by physicists active at the time. Dean Rickles and Donald Salisbury, interview with Louis Witten, 17 March 2011, https://www.aip.org/history-programs/niels-bohr-library/oral-histories/36985. Accessed 12 March 2017.

attempt to build an international community of "relativists" and all the problematic aspects that this community building and institutionalization process entailed in that particular historical period. As briefly mentioned in Chap. 1, I focus mainly on two elements. The first concerns the unsettled epistemic status of the theory at the time. The socio-epistemic dispersion identified in our description of the low-water-mark period was still ongoing in the 1950s and was one of the major obstacles to overcome in the attempts to build a community. The epistemic dispersion characterizing the relationships between the different, loosely connected, research agendas as well as national and disciplinary divides still shaped the work carried out at the research centers. This implied that it was not easy to envisage a common framework from the different research activities and that the attempts to do so had consequences on the research activities themselves. The second element regarded the problem of how to structure a community in the international arena during the Cold War. At the time, community builders could look only to a very few examples of organized international scientific collaboration, and these structures imposed constraints on the ways in which institutional community building was to be pursued. As we shall see, these constraints allowed the actors to initiate the process in the first place but also created a series of problems when politics suddenly entered the equation. In the next chapter, I will discuss the existing structures of international scientific collaboration that served as models for the construction and institutionalization of the GRG community. These were existing institutional bodies that, although created before World War II, were being transformed in the changing political climate of the postwar and Cold War periods.

References

Bettencourt, Luís M.A., David I. Kaiser, Jasleen Kaur, Carlos Castillo-Chávez, and David E. Wojick. 2008. Population modeling of the emergence and development of scientific fields. *Scientometrics* 75: 495–518. doi:10.1007/s11192-007-1888-4.

Blum, Alexander, Roberto Lalli, and Jürgen Renn. 2015. The reinvention of general relativity: A historiographical framework for assessing one hundred years of curved space-time. *Isis* 106: 598–620.

Blum, Alexander, Roberto Lalli, and Jürgen Renn. 2016. The renaissance of general relativity: How and why it happened. *Annalen der Physik* 528: 344–349. doi:10.1002/andp.201600105.

Blum, Alexander, Domenico Giulini, Roberto Lalli, and Jürgen Renn. 2017. Editorial introduction to the special issue "The Renaissance of Einstein's Theory of Gravitation". *The European Physical Journal H* 42: 95–105. doi:10.1140/epjh/e2017-80023-3.

Blum, Alexander, and Dean Rickles (eds.). 2017. *Quantum gravity in the first half of the twentieth century: A sourcebook.* Berlin: Edition Open Access.

Bonolis, Luisa. 2017. Stellar structure and compact objects before 1940: Towards relativistic astrophysics. *The European Physical Journal H* 42: 311–393. doi:10.1140/epjh/e2017-80014-4.

Cartan, Élie. 1922a. Sur les espaces généralisés et la théorie de la Relativité. *Comptes Rendus* 174: 734–737.

Cartan, Élie. 1922b. Sur les espaces conformes généralisés et l'Univers optique. *Comptes Rendus* 174: 857–860.

van Dongen, Jeroen. 2010. *Einstein's unification*. Cambridge: Cambridge University Press.

Eisenstaedt, Jean. 1986. La relativité générale à l'étiage: 1925–1955. *Archive for History of Exact Sciences* 35: 115–185.

Eisenstaedt, J. 1987. Trajectoires et impasses de la solution de Schwarzschild. *Archive for History of Exact Sciences* 37: 275–357.

Eisenstaedt, Jean. 1989. The low water mark of general relativity, 1925–1955. In *Einstein and the history of general relativity*, ed. Don Howard and John Stachel, 277–292. Boston: Birkhäuser.

Eisenstaedt, Jean. 2006. *The curious history of relativity: How Einstein's theory of gravity was lost and found again*. Princeton: Princeton University Press.

Ellis, George. 2012. Editorial note to: H. P. Robertson, Relativistic cosmology. *General Relativity and Gravitation* 44: 2099–2114. doi:10.1007/s10714-012-1400-1.

Goenner, Hubert. 2004. On the history of unified field theories. *Living Reviews in Relativity* 7: 2. doi:10.12942/lrr-2004-2.

Goenner, Hubert. 2014. On the history of unified field theories, part II. (ca. 1930–ca. 1965). *Living Reviews in Relativity* 17: 5. doi:10.12942/lrr-2014-5.

Goenner, Hubert. 2017. A golden age of general relativity? Some remarks on the history of general relativity. *General Relativity and Gravitation* 49: 42. doi:10.1007/s10714-017-2203-1.

Goldstein, Catherine, and Jim Ritter. 2003. The varieties of unities: Sounding unified theories 1920–1930. In *Revisiting the foundations of relativistic physics: Festschrift in honor of John Stachel*, ed. Abhay Ashtekar, Robert S. Cohen, Don Howard, Jürgen Renn, Sahotra Sarkar, and Abner Shimony, 93–149. Dordrecht: Kluwer.

Gutfreund, Hanoch, and Jürgen Renn. 2017. *The formative years of relativity: The history and meaning of Einstein's Princeton lectures*. Princeton: Princeton University Press.

Jordan, Pascual. 1955. *Schwerkraft und Weltall: Grundlagen der theoretischen Kosmologie. Wissenschaft, Bd. 107*. Braunschweig: F. Vieweg.

Kaiser, David. 2000. Roger Babson and the rediscovery of general relativity. In *Making theory: Producing theory and theorists in postwar America,* 567–595. Ph.D. dissertation, Harvard University.

Kaiser, David. 2005. *Drawing theories apart: The dispersion of Feynman diagrams in postwar physics*. Chicago: University of Chicago Press.

Kaiser, David. 2012. Booms, busts, and the world of ideas: Enrollment pressures and the challenge of specialization. *Osiris* 27: 276–302. doi:10.1086/667831.

Kennefick, Daniel. 2007. *Traveling at the speed of thought: Einstein and the quest for gravitational waves*. Princeton: Princeton University Press.

Kragh, Helge. 1999. *Quantum generations: A history of physics in the twentieth century*. Princeton: Princeton University Press.

Kragh, Helge, and Dominique Lambert. 2007. The Context of discovery: Lemaître and the origin of the primeval-atom universe. *Annals of Science* 64: 445–470. doi: 10.1080/00033790701317692.

Kruskal, Martin D. 1960. Maximal extension of Schwarzschild metric. *Physical Review* 119: 1743–1745. doi: 10.1103/PhysRev.119.1743.

Kuhn, Thomas S. 1970. *The structure of scientific revolutions*, 2nd ed. Chicago: Chicago University Press.

Longair, Malcolm S. 2006. *The cosmic century: A history of astrophysics and cosmology*. Cambridge, UK: Cambridge University Press.

Newman, Ezra, and Roger Penrose. 1962. An approach to gravitational radiation by a method of spin coefficients. *Journal of Mathematical Physics* 3: 566–578. doi:10.1063/1.1724257.

Ortega-Rodríguez, M., H. Solís-Sánchez, E. Boza-Oviedo, K. Chaves-Cruz, M. Guevara-Bertsch, M. Quirós-Rojas, S. Vargas-Hernández, and A. Venegas-Li. 2017. The early scientific contributions of J. Robert Oppenheimer: Why did the scientific community miss the black hole opportunity? *Physics in Perspective* 19: 60–75. doi:10.1007/s00016-017-0195-6.

Peebles, Phillip James Edwin. 2017. Robert Dicke and the naissance of experimental gravity physics, 1957–1967. *The European Physical Journal H* 42: 177–259. doi:10.1140/epjh/e2016-70034-0.

Penrose, Roger. 1960. A spinor approach to general relativity. *Annals of Physics* 10: 171–201. doi:10.1016/0003-4916(60)90021-X.

Petrov, Alekei Z. 2000. The classification of spaces defining gravitational fields. *General Relativity and Gravitation* 32: 1665–1685. doi: 10.1023/A:1001910908054.

Renn, Jürgen (ed.). 2007. *The genesis of general relativity, 4 Vols*. Dordrecht: Springer.

Robinson, Ivor, Alfred Schild, and E. L. Schucking (eds.). 1965. *Quasi-stellar sources and gravitational collapse, including the proceedings of the First Texas Symposium on relativistic astrophysics*. Chicago: University of Chicago Press.

Rickles, Dean. 2015. Institute of Field Physics, Inc: Private Patronage and The Renaissance of Gravitational Physics. Talk presented at the conference *A Century of General Relativity*, Berlin, 4 December 2015.

Szekeres, George. 1960. On the singularities of a Riemannian manifold. *Publicationes Mathematicae Debrecen* 7: 285–301.

Thorne, Kip S. 1994. *Black holes and time warps: Einstein's outrageous legacy*. New York: WWNorton.

Vizgin, V. P., and G. E. Gorelik. 1987. The reception of the theory of relativity in Russia and the USSR. In *The comparative reception of relativity*, ed. Thomas F. Glick, 354–363. Dordrecht: Reidel.

Will, Clifford. 1986. *Was Einstein right?: Putting general relativity to the test*. New York: Basic Books.

Will, Clifford. 1989. The renaissance of general relativity. In *The new physics*, ed. Paul Davies, 7–33. Cambridge: Cambridge University Press.

Wright, Aaron Sidney. 2014. The advantages of bringing infinity to a finite place. *Historical Studies in the Natural Sciences* 44: 99–139. doi:10.1525/hsns.2014.44.2.99.

Chapter 3
(Re-)Establishing International Cooperation After World War II

Abstract The complex landscape of the scientific institutions operating at the international level in the post-World War II period is outlined here. Around the mid-1950s, when the community-building activities connected to general relativity first began, a reconfiguration of these institutions for the promotion and organization of international cooperation in science was under way. The motivations for, and constraints of, this transformation were defined by the world order that was being constructed after the end of World War II and by the evolution of the Cold War. For those willing to create a new structure for promoting general relativity in the international arena, these existing institutions provided both a model to follow and a larger established structure with which to interact. It is argued that one of the major structural changes in institutions such as the International Unions was that they began promoting specific areas of research at this point, while before World War II their role was limited to define international standards. Besides these structural changes in scientific institutions, the second major element was the changing political context related to the post-Stalinist reforms in the Soviet Union and the related détente in international relations that led to an increasing participation of Soviet scientists in international scientific institutions.

Keywords Cold war · International Council of Scientific Unions · International relations · International Union of Pure and Applied Physics · Scientific internationalism · Scientific institutions · Soviet Union · UNESCO

The idea that science is intrinsically universal has permeated the meta-scientific discourse on the subject for centuries. Whether or not this ideal has been translated into practices informed by cosmopolitan principles such as free circulation of ideas and persons across boundaries—national or of other kinds—invariably depended on the historical contexts. The rise of modern nation-states and the increasing involvement of scientists in affairs of national relevance have created a fundamental contradiction between this cosmopolitan ideal and daily scientific practices. Since the 17th century, universalistic principles in the world of science have had to face concerns related to national interests. Which direction prevailed among scientists

R. Lalli, *Building the General Relativity and Gravitation Community During the Cold War*, SpringerBriefs in History of Science and Technology,
DOI 10.1007/978-3-319-54654-4_3

largely depended on the developments of national politics and the associated international relations (see, e.g., Forman 1973; Schroeder-Gudehus 1978, 1990; Fox 2016).

In the history of the dynamic contrast between internationalism and nationalism in science, a crucial development occurred in the second part of the 19th century. Paradoxically, when state apparatuses started to consistently perceive scientific research as a necessary element of national well-being in terms of industrial and economic developments, scientists began organizing themselves into institutional bodies that promoted scientific collaborations between scientists and institutions located in different countries (Crawford 1990). These initial attempts at institutionalizing international cooperation were shaken by dramatic political events during the 20th century. Some of these early institutions succumbed but the majority showed great resilience and transformed into new structures better suited to the changing political environment. By the time of the events addressed in this book, a re-configuration of these institutional structures was under way. The motivations and constraints of this transformation were defined by the world order that was being constructed after the end of World War II and by the evolution of the Cold War. For those willing to create new institutional bodies, these existing institutions provided both a model to follow and a larger established structure with which to interact.

After World War II, the main and largest of these institutions for furthering international scientific cooperation was the International Council of Scientific Unions (ICSU), which still exists to this day under the name of International Council for Science. Established in 1931, ICSU is the successor body of the International Research Council (1919–1931), in turn the successor body of the International Association of Academies (1899–1914). Like the International Research Council before it, ICSU is a non-governmental confederation of national academies and international unions devoted to specific scientific disciplines. From an organizational perspective, the major difference between ICSU and the International Research Council was that, after the 1931 reform, the international scientific unions had much more representative and decision-making power. In the immediate aftermath of World War II, there were seven of these international unions: the International Astronomical Union (IAU), the International Union of Biological Sciences (IUBS), the International Union of Geodesy and Geophysics (IUGG), the International Union of Geological Sciences (IUGS), the International Union of Pure and Applied Chemistry (IUPAC), the International Union of Pure and Applied Physics (IUPAP), and the International Union of Radio Science (URSI). Within the space of a few years, they would grow in number and include other scientific disciplines. One of these was the newly re-established International Mathematical Union (IMU), which was re-admitted to ICSU in 1952.

The foundation of the United Nations in the immediate aftermath of World War II shaped the institutional reconstruction of international cooperation by providing a more solid institutional basis for international scientific exchange through the establishment of the United Nations Educational Scientific Cultural Organization (UNESCO). Recognizing the role of ICSU as a well-established body for the promotion of science, UNESCO officials resolved to substantially assist ICSU with financial and staff support. In December 1946, UNESCO signed a formal agreement with ICSU in which it was determined that the two organizations would act in a coordinated manner to establish and promote international collaboration and exchange in science (Sewell 1975, pp. 94–96; Greenaway 1996; Lehto 1998). From that moment onward, one of ICSU's mandates has been to allocate UNESCO funds to scientific unions, which also made the unions much more dependent on ICSU's support for pursuing their activities (see, e.g., Fennell 1994, p. 113; Lehto 1998, p. 100).

The establishment of UNESCO led to a radical transformation of how international scientific cooperation was organized at the institutional level. According to historians of science Elisabeth Crawford, Terry Shinn, and Sverker Sörlin, it is possible to categorize two different modes of organization of international scientific institutions: *spontaneous* and *bureaucratic* (Crawford et al. 1993, pp. 23–25). Under the first mode, Crawford and her co-authors include all the different types of organizations founded in the late 19th century or the first half of the 20th century. This mode is defined by the fact that "the moving force is the interests of individual scientists who draw on national resources to hold world congresses, set up committees on standardization or coordinate national projects" (p. 23). The spontaneous mode of operation neither had the capacity nor aimed to undertake more ambitious projects such as setting up transnational research programs and institutions. ICSU and its unions were the most important examples of this mode of spontaneous organization. Within this perspective, the historical evolution of these entities toward more organized structures by the late 1930s shows how this modality was able to achieve a better level of coordination without loss of spontaneity. The second mode, the bureaucratic one, was a product of the post-World War II period. It was predominantly embodied by UNESCO, which was established with the more ambitious goal of setting transnational scientific agendas with a stronger bureaucratic apparatus.

Although the two-mode categorization described above is certainly useful for framing the historical trajectory of the institutionalization of scientific cooperation in the international arena, my account will depart from it in two important aspects. First, since their inception in the interwar period, ICSU and the scientific unions had institutional constraints that made them appear rather bureaucratic compared to other forms of organized scientific cooperation, and it was precisely these bureaucratic regulations that were perceived as essential by emerging institutions that saw ICSU and the international unions as role models. The most important point was that participation in the governing bodies of ICSU and its international unions was based on the notion of "national membership." In 1931, when ICSU was established, two articles regulated the organization's membership rules:

1. The ICSU consists of a national scientific organization from each country which has adhered to the Council and of the International Unions.
2. A country may join the ICSU either through its principal Academy, or through its national Research Council, or through some other national institution or

association of institutions, or, in absence of these, through its Government. (Greenaway 1996, p. 190).

These two articles and the definition of national membership did not change significantly after the 1952 revision of ICSU's statute (Greenaway 1996, p. 190). Although the structures of the international unions affiliated with ICSU followed very specific regulations, all of them employed a similar form of membership in which each country was represented by national delegates.[1] During the Cold War, these regulations were to play a fundamental role as the meanings and functions of national academies and national delegates depended on which side of the Iron Curtain these entities were on. In liberal democracies, national academies and research councils normally had different functions related to authority allocation and to research funding, respectively. By contrast, in centrally organized socialist countries, national academies generally had a much broader role in governing and regulating the scientific enterprise by allocating both resources and status. Moreover, in Eastern Bloc countries, academies of sciences depended more on political patronage (Rabkin 1988, p. 17; Hall 2003).

The second aspect which is not captured by the ICSU-UNESCO dichotomy as representation of the spontaneous versus bureaucratic mode of organization but is fundamental to the argument of the present work is that the functions and organization of ICSU itself, and of its unions, considerably changed after World War II. Partly as a consequence of the formal and financial ties with UNESCO, ICSU and the unions became more bureaucratic in two fundamental, interconnected ways. During the 1950s, in relation to the shift of UNESCO policies toward the planning of major projects, ICSU itself became more project-oriented (Sewell 1975, p. 120). ICSU began to act as the major organizing body of ambitious international projects, the most relevant of which was the International Geophysical Year (IGY) in 1957–1958 (Greenaway 1996, pp. 91–92). More important to my argument, the definition of "project" used by UNESCO officials was broad enough to cover the creation of new institutional bodies devoted to specific areas of research (Greenaway 1996, p. 85). And this special feature re-defined even more markedly the nature of ICSU and its unions. In the aftermath of World War II, new branches of scientific research mushroomed as a consequence of the exponential growth of science. In principle, this explosion of scientific activities could have led to the formation of brand new unions devoted to the emerging specializations to be later incorporated into ICSU. In order to better organize the allocation of funds provided by UNESCO, however, ICSU preferred to admit as members only larger unions devoted to encompassing disciplines such as physics, chemistry, mathematics, etc., with the different unions grouping together different sub-disciplines. The organizational style that was envisaged as soon as the change in scale of the scientific enterprise became evident was pyramidal, with ICSU at the top, the larger unions and ICSU committees in the middle, and smaller sections of the unions at the lower level (see, e.g., Fennell

[1]When IMU was re-established in 1952, for example, the matter was discussed and it was decided that membership was to be limited to countries (Lehto 1998, p. 98).

1994, pp. 113–117; Greenaway 1996, pp. 85–86).[2] This change, which might at a first sight seem purely organizational, had a strong impact on the role of unions and on their range of activities.

For the scope of this book, it is important to understand how these programmatic reconfigurations were implemented within the International Union of Pure and Applied Physics (IUPAP), which was to have a predominant role in the activities of the International Committee on General Relativity and Gravitation (ICGRG) and the establishment of the International Society on General Relativity and Gravitation (ISGRG). Before World War II, IUPAP's commissions dealt solely with general subjects mostly related to the definition of internationally shared standards. This is not surprising. Historians of science have convincingly argued that the major force leading to the strengthening of international cooperation and the founding of international institutions was the need to establish common standards at the cognitive, communicative, and technical levels (Crawford et al. 1993, pp. 11–19). Besides the Commission on Finances, the other two IUPAP commissions launched prior to 1939 dealt with topics such as "Publications" and "Symbols Units and Nomenclature." Both these commissions had clearly been intended to work on the different kinds of standardization procedures previously mentioned, with the explicit goal of facilitating communication between scholars working in different national contexts.

Soon after the end of World War II, IUPAP commissions dedicated to specific disciplinary fields were established for the first time, with the first two being on thermodynamics and statistical mechanics in 1945, and on cosmic rays in 1947. The goal of these commissions was mixed. In part, these commissions were created in order to deal with concerns about different standards in specific research areas; namely, they continued to have the standardization function of pre-World War II commissions, but now limited to a specific research topic. More subtly, by promoting scientific cooperation across national boundaries on specific scientific sub-fields, the commissions served the purpose of strengthening research in these particular fields as much as, or more than, creating a common background for the different national scientific traditions (Anon. 1992).

The second important, and related, structural change to IUPAP concerned the official recognition of a different kind of commission. At the fifth General Assembly in 1947, it was decided to admit Affiliated Commissions. These commissions were similar in scope to the internal commissions dedicated to specific fields established by IUPAP at around the same time. The first—and, for the following sixteen years, the only—Affiliated Commission was the International Commission for Optics, established in 1948. For this commission, the rule of national membership continued to hold. The members of the International Commission for Optics were countries that sent their delegates to the meetings of the commission. The only

[2]The major exception to this unwritten rule was in biology, when new unions in specialized fields of biology were founded and admitted: the International Union of Immunological Societies in 1976 and the International Union of Microbiological Societies in 1982 (see Greenaway 1996, pp. 128-131).

major difference between Affiliated Commissions and internal commissions was that the Affiliated Commission was a body comprising groups of scientists recognized by the IUPAP General Assembly, but not appointed by IUPAP itself (Anon. 1992; Howard 2003). The changing function of the internal commissions and the new form of Affiliated Commissions both denote a gradual, but deep, transformation in the function of IUPAP, which would become increasingly evident over the years. In the postwar period, the organization redefined itself by establishing institutionalized forms of international cooperation aimed at developing *specific fields*—an activity that was simply not pursued in the pre-World War II period.

These kinds of transformation of the committees' role were not specific to IUPAP. ICSU itself began setting up specific committees on emerging fields such as the Committee on Space Research (COSPAR) and the Scientific Committee on Antarctic Research (SCAR), both established in 1958 as by-products of the IGY and its success (Greenaway 1996, p. 157). Within the unions, a similar pattern is evident. A striking example is provided by the structural developments of the International Union of Pure and Applied Chemistry (IUPAC). At the first IUPAC conference held in Rome in 1920, the permanent commissions established by the General Assembly dealt with the topics of atomic weights, tables of constants, chemical standards, and patents, and it was decided to establish a commission on nomenclature the following year. Subsequently, the number of permanent commissions grew considerably, but all of them dealt specifically with issues of epistemic, communicative and technical standards, plus terminology and hygienic norms. In 1951, the IUPAC established six different sections, each devoted to a particular subfield (physical chemistry, inorganic chemistry, organic chemistry, biological chemistry, analytical chemistry, and applied chemistry). While, initially, the existing commissions dealing with standardization procedures and terminology matters were simply relocated to a specific section, this seemingly purely organizational renovation came to imply that the nature of the union changed completely. The very first rule of the by-laws drafted in 1951 read that the function of the section committees was "to organize Conferences, Congresses, Discussions and other meetings considered useful for the developments of the scientific and technical field covered by the Section"—a much broader scope than simply providing an arena for discussions and agreements about standardization and terminology issues (Fennell 1994, p. 116). In the subsequent reorganization of the individual sections, this change of role became even more explicit when new commissions devoted to specific areas of research were appointed, with exactly the same phenomenon occurring at IUPAP at about the same time. This was particularly evident in the Physical Chemistry Section, within which six new commissions on specific research areas were soon established.[3]

[3]The six research sub-fields were: macromolecules, radioactivity, electrochemistry, chemical thermodynamics, molecular spectroscopy, and kinetics of chemical reactions (Fennell 1994, p. 119).

So far, I have only discussed structural aspects of the relevant institutional frameworks, their main ingredients, and their organizational transformations, which, combined, constituted the backdrop of the institutionalization process of the GRG community that was to enter IUPAP in 1975 as the Second Affiliated Commission. As I have attempted to show, organizational changes did not have purely logistic significance. They had effects on the functions performed by these bodies and, consequently, on their very nature. Important as they are, structures and functions, however, were not the only thing that changed in the post-World War II period. Other relevant transformations occurred, and they were mostly of a political nature.

To begin with, UNESCO's mission, policies, and range of actions were historically contingent. The balance between universalistic principles and *realpolitik* considerations was one of the elements more dependent on political events and changes of leadership. While the early stage was characterized by a high degree of ideality with the first two general directors willing to keep the non-governmental body independent of the influences of national governments in the pursuit of the UNESCO mission, by the late 1940s, it was evident that national political pressures were increasingly shaping its activities. When American political scientist Luther H. Evans became Director General of UNESCO in 1953, he explicitly recognized that it was a fact that national governments played a predominant role in the UNESCO decision-making process and that it was detrimental to challenge this state of affairs (Sewell 1975, p. 166). Evans believed that the increasing commitment of member states was a condition *sine qua non* in order for UNESCO international efforts to be effective. All through the 1950s, his diplomatic management style was shaped by this standpoint.

Evans' honest assessment of the contradiction between the inter-governmental UNESCO missions and the national interests of the state members that made it work, and his subsequent inclination toward *realpolitik* was part of the largest framework for international institutional cooperation after the mid-1950s, when the first steps toward the formation of the GRG community were taken. Both the recognition of the problem and Evans' policies set an example for community builders. Through explicit assessment of the intrinsic cosmopolitanism-nationalism dichotomy within the framework provided by UNESCO, Evans made the terms of the problem a subject of debate between the officials of institutional bodies devoted to further international collaboration in science. More importantly, his decision provided a model of diplomatic goal-directed actions, which were characterized by the renunciation of some elements of the idealistic principles underlying the goals themselves. The Director General of UNESCO was thus creating a role model of a realistic, diplomatic scientist-administrator during the Cold War.

The second relevant political element concerned developments of international relations and, more specifically, of East-West relations. All the activities of re-establishing, or establishing, international cooperation were shaped by the changing political climate of the Cold War. In particular, attempts to form international links in research fields connected to general relativity began in a period when a relaxation in East-West relations was made possible by the end of the Korean conflict and by post-Stalinist reforms in the Soviet Union. A few years after the death of Joseph Stalin in March 1953, Soviet rulers decided that the recent evolution of nuclear weaponry and its destructive power must be balanced with a policy of "peaceful coexistence" between different economic and political systems. This novel Soviet foreign policy—made official by Nikita Khrushchev during the 20th Congress of the Communist Party of the Soviet Union in 1956—led to a relaxation of international tensions and dramatically increased the opportunities for Soviet scientists to participate in scientific exchanges with colleagues working in Western countries (Graham 1993; Holloway 1994; Mastny 2010; Hollings 2016).

The likelihood of establishing international ties had been low since the mid-1930s, for scientific activities were carried out under tight bureaucratic control and with the constraint of being consistent with the official ideology of the party, the Marxist philosophy of dialectical materialism (Graham 1972; Pollock 2006).[4] In the early Cold War period, contacts with Western scientists remained rare, and the participation of Soviet scientists in international institutions was almost nonexistent. One example that illustrates this is UNESCO. Whereas the Soviet Union played a pivotal role in the foundation of the United Nations and took part in the discussions which eventually led to the establishment of UNESCO, Soviets did not join UNESCO until the change in political leadership (Armstrong 1954). After this change occurred, the post-Stalinist reforms in foreign and domestic policies resulted in a tremendously rapid increase in Soviet membership of international scientific institutions.[5] In 1954, the Soviet Union also became a member of ICSU through the affiliation of the Academy of Sciences of the USSR, which paved the way for its affiliation to IMU and IUPAP in 1957, its participation in the IGY project in 1957–1958, and the establishment of COSPAR in 1958, to name but a few. One of the effects of increasing Soviet involvement in international bodies for scientific cooperation was a higher level of involvement of other countries under its sphere of influence. An indication is given again by the increasing participation of Eastern Bloc countries in UNESCO's activities. Shortly after the Soviet Union joined UNESCO, Czechoslovakia, Hungary, and Poland began to attend its meetings again after a multi-year period of disengagement. Later, Bulgaria, Rumania, and Albania became members, the first two in 1956 and the latter in 1958 (Sewell 1975, p. 168).

The IGY had a particularly strong relevance in shaping the future evolution of international scientific cooperation as far as East-West, and more specifically American-Soviet, scientific exchanges are concerned. The successful Soviet launch in October 1957 of the first artificial Earth satellite, Sputnik 1, was taken as proof of how advanced Soviet technological and scientific capabilities were. Besides opening up a new scientific field—space science—the Sputnik shock had tremendous consequences on domestic and foreign science policies in the United States.

---

[4]For the impact on physics research, see Josephson (1991) and Kojevnikov (2004). For the specific case of research on general relativity, see Vizgin and Gorelik (1987).

[5]According to Russian historian of science Konstantin Ivanov, in the post-World War II period the USSR was a member of only two international scientific institutions before 1953 (Ivanov 2002, p. 321), one of which was certainly the IAU, of which it had been a member since 1935 (Hollings 2016, p. 31).

Domestically, it led to a dramatic increase in both scientific manpower and facilities supported by the federal government under the strong impression that the United States was lagging behind the Soviet Union in this rapidly developing, and military relevant, scientific area (Kaiser 2002). In the international arena, it resulted in a rapid expansion of bilateral exchanges between Soviet and American scientists, with the latter interested in gaining information on the Soviet scientific and engineering achievements (Rabkin 1988, p. 12).

East-West scientific relations initiated during Khrushchev's thaw and increased during the IGY were characterized by a mixture of political, scientific, and moral agendas, both in individuals' choices and governments' actions. The development of Soviet-American scientific exchanges after the Sputnik shock was a case in point. Governments were interested in promoting international scientific exchanges for a variety of political reasons: appropriating the other's technological and scientific advances, using the need for interchange as a bargaining chip in foreign policy, or exploiting scientists' close collaboration as a tool of ideological propaganda. The combination of reasons was of course very different for the USSR and for the United States, and, at the same time, historically dependent (Rabkin 1988, p. 44). Scientists had their own agenda, too. They were interested in pursuing exchanges with their peers working on the other side of the Iron Curtain for scientific reasons, of course, but also for more mundane or political purposes. On the one hand, international scientific recognition could increase personal prestige at home and further an individual's career. On the other hand, many scientists considered Soviet-American scientific exchanges to also be political tools, albeit used for quite different ends. One major argument, used by physicists in particular, was that these collaborations were a necessary step toward world peace and nuclear disarmament. Many, however, stuck to the agendas promoted by their governments and came to see, implicitly or explicitly, international exchanges as political tools in the ways addressed by those governments.[6] This complex variety of scientific and political reasons produced a changing model of "state-sponsored," state-supported scientific internationalism. Scientists' individual goals, which might have been motivated by genuine scientific interests or by the idealistic desire to help the world peace process, had to be adjusted to deal with this.[7]

In the Soviet Union, the recovery of international contacts during Khrushchev's thaw remained strongly bureaucratized with the implementation of rigid regulations concerning how Soviet scientists should obtain permission to travel abroad or invite foreign scholars, or even to maintain correspondence with foreign institutions and persons (Gerovitch 2002, pp. 156–157). This bureaucratized, state-controlled style of organizing scientific exchanges shaped the process in several ways. Not least, it produced a highly regulated system that was accepted by other countries wishing to

[6]One way to look at this agglomeration of scientific, personal, and political reasons, at least in the adopted rhetoric, is Aant Elzinga's distinction between "autoletic scientific nationalism" and "heteroletic scientific nationalism" (Elzinga 1996, p. 38).

[7]On the notion of state-sponsored internationalism, see Wang (1999).

continue collaborating. In the United States, agencies were created de novo that dealt uniquely with the process of scientific exchanges with the Soviet Union. In their practices, these agencies came to mimic the centralized organizational mode typical of the Soviet political-scientific apparatuses (Rabkin 1988, pp. 25–37). While these features have traditionally been considered a concern for American–Soviet relations, and as such they had been a subject of analysis, they hardly remained confined to scientific relations between the two superpowers. The same type of concerns, agendas, and structures of collaboration had a relevant place in the whole re-establishment of East-West scientific relations in the Cold War period, and were of particular relevance in the institutional configurations created to promote international cooperation worldwide.

Overall, the regulation of national membership dictating scientists' participation in institutional bodies such as ICSU and its unions was an essential element in furthering East-West scientific collaborations. This suited the political and cultural structures governing the relationships between policy and science in the Soviet Union well. Membership of these bodies through the Academy of Sciences of the USSR allowed Soviet political leaders to exercise direct control in the choice of national delegates and in their activities. In the changing political climate following Stalin's death, these structures based on national membership facilitated the inclusion of Soviet scientists/delegates in international scientific institutions since they did not jeopardize the centralized management style of the Soviet political system. Because of the guarantee their structures offered for re-establishing international cooperation in the polarized Cold War from the 1950s onward, and because of their changing functions described above, ICSU, the international unions, and their commissions became the institutional frameworks in which scientists normally attempted to promote their disciplines and sub-disciplines in the international arena. The institutionalization process of the community of scholars working in fields related to GRG followed a different trajectory, which was shaped by the epistemic features of the scientific field under construction as well as by the political developments of the Cold War.

References

Anon. 1992. *UIPPA-IUPAP 1922–1992*. Album souvenir realized in Quebec by the Secretariat of IUPAP. http://iupap.org/wp-content/uploads/2013/04/history.pdf. Accessed 7 Mar 2016.

Armstrong, John A. 1954. The Soviet attitude toward UNESCO. *International Organization* 8: 217–233.

Crawford, Elisabeth. 1990. The universe of international science, 1880–1939. In *Solomon's house revisited: The organization and institutionalization of science*, ed. Tore Frängsmyr, 251–269. Canton, MA: Science History Publications.

Crawford, Elisabeth, Terry Shinn, and Sverker Sörlin. 1993. The nationalization and denationalization of the sciences: An introductory essay. In *Denationalizing Science*, ed. Elisabeth Crawford, Terry Shinn, and Sverker Sörlin, 1–42. Dordrecht: Springer. doi: 10.1007/978-94-017-1221-7_1.

Elzinga, Aant. 1996. Modes of internationalism. In *Internationalism and science*, ed. Aant Elzinga, and Catharina Landstrom, 3–20. London: Taylor Graham.

Fennell, Roger. 1994. *History of IUPAC, 1919–1987*. Oxford: Blackwell Science Ltd.

Forman, Paul. 1973. Scientific internationalism and the Weimar physicists: The ideology and its manipulation in Germany after World War I. *Isis* 64: 151–180.

Fox, Robert. 2016. *Science without frontiers: Cosmopolitanism and national interests in the world of learning, 1870–1940*. Corvallis, OR: Oregon State University Press.

Gerovitch, Slava. 2002. *From Newspeak to cyberspeak: A history of soviet cybernetics*. Cambridge, MA: MIT Press.

Graham, Loren R. 1972. *Science and philosophy in the Soviet Union*. New York: Knopf.

Graham, Loren R. 1993. *Science in Russia and the Soviet Union: A short history*. Cambridge: Cambridge University Press.

Greenaway, Frank. 1996. *Science international: A history of the International Council of Scientific Unions*. Cambridge: Cambridge University Press.

Hall, Karl. 2003. Europe and Russia. In *The Oxford companion to the history of modern science*, ed. John Heilbron, 279–282. Oxford: Oxford University Press.

Hollings, Christopher D. 2016. *Scientific communication across the Iron Curtain*. Cham: Springer International Publishing. doi: 10.1007/978-3-319-25346-6.

Holloway, David. 1994. *Stalin and the bomb: The Soviet Union and atomic energy, 1939–1956*. New Haven: Yale University Press.

Howard, John N. 2203. The early meeting of the International Commission of Optics. *Optics and Photonics news*. June 2003: 16–17.

Ivanov, Konstantin. 2002. Science after Stalin: Forging a new image of Soviet science. *Science in Context* 15: 317–338.

Josephson, Paul R. 1991. *Physics and politics in revolutionary Russia*. Berkeley: University of California Press.

Kaiser, David. 2002. Cold War requisitions, scientific manpower, and the production of American physicists after World War II. *Historical Studies in the Physical and Biological Sciences* 33: 131–159. doi:10.1525/hsps.2002.33.1.131.

Kojevnikov, Aleksei B. 2004. *Stalin's great science: The times and adventures of Soviet physicists*. London: Imperial College Press.

Lehto, Olli. 1998. *Mathematics without borders: A history of the International Mathematical Union*. New York: Springer.

Mastny, Vojtech. 2010. Soviet foreign policy, 1953–1962. In *The Cambridge History of the Cold War, Vol. 1, Origins*, ed. Melvyn P. Leffler, and Odd A. Westad, 312–333. Cambridge: Cambridge University Press.

Pollock, Ethan. 2006. *Stalin and the Soviet science wars*. Princeton: Princeton University Press.

Rabkin, Yakov M. 1988. *Science between the superpowers*. New York: Priority Press.

Schroeder-Gudehus, Brigitte. 1978. *Les scientifiques et la paix: La communauté scientifique internationale au cours des années 20*. Montréal: Presses de l'Université de Montréal.

Schroeder-Gudehus, Brigitte. 1990. Nationalism and internationalism. In *Companion to the History of Modern Science*, ed. R. C. Olby, G. N. Cantor, J. R. R. Christie, and M. J. S. Hodge, 909–19. Routledge.

Sewell, James Patrick. 1975. *UNESCO and world politics: Engaging in international relations*. Princeton, N.J.: Princeton University Press.

Vizgin, V.P., and G.E. Gorelik. 1987. The reception of the theory of relativity in Russia and the USSR. In *The comparative reception of relativity*, ed. Thomas F. Glick, 354–363. Dordrecht: Reidel.

Wang, Zuoyue. 1999. U.S.-China scientific exchange: A case study of state-sponsored scientific internationalism during the Cold War and beyond. *Historical Studies in the Physical and Biological Sciences* 30: 249–277. doi: 10.2307/27757826.

Chapter 4
The Formative Phase of the GRG Community

Abstract This chapter analyzes the first decade of the process of building the community of "relativists" from the mid-1950s to the mid-1960s. This period can be seen as the formative phase of the emerging community, during which the initial steps were taken to institutionally unify the different research agendas under the heading of "General Relativity and Gravitation." These included the organization of the first international conference on general relativity held in Bern in 1955, the establishment of the International Committee on General Relativity in Gravitation in 1959, and the decision to publish the *Bulletin on General Relativity and Gravitation* from 1962 onward. It is argued that some of the initial impetus to build the international community was related to idealistic views about the role of science in achieving peaceful relations between nations. By the end of the formative phase, however, many tensions of both a political and epistemic nature came to dominate the discussions about the future of the committee as it had to face sudden changes in the social composition of the community as well as in the redefinition of the field after the discovery of quasars in 1963 and the emergence of relativistic astrophysics.

Keywords Albert Einstein · André Mercier · Bulletin on General Relativity and Gravitation · Cold War · Community building · International Committee on General Relativity and Gravitation · International Conferences on General Relativity and Gravitation · International Union of Pure and Applied Physics · Relativistic astrophysics · Renaissance of general relativity

In Chap. 2, I argued that the landscape of research in the mid-1950s on topics that would soon be considered part of the GRG domain was characterized by a dispersed set of small research centers. One central argument of the present work is that the community-building activities beginning around the mid-1950s had a pivotal role in, and were a fundamental component of, the renaissance phenomenon. The reason was that these social activities helped bridge the work carried out at these different centers, by strengthening, or creating for the first time, connections between the various researchers interested in general relativity. This

R. Lalli, *Building the General Relativity and Gravitation Community During the Cold War*, SpringerBriefs in History of Science and Technology,
DOI 10.1007/978-3-319-54654-4_4

function seems to be confirmed by scientists who evaluated the status of the field before these activities commenced. In later recollections, some actors in fact considered the dispersion of the field one of the major causes of its stagnation during the low-water-mark period.

This state of affairs is confirmed by the fact that there was no clear perception of the overall status of the field in the mid-1950s. As described in detail in Appendix A, in 1955, there was a whole host of centers that were pursuing, or in the process of establishing, a research agenda related to general relativity. In various countries, these scientific activities were growing, albeit slowly. However, this increasing activity was not clearly perceived by the actors at the time. In a public speech delivered in December 1955 as Director of the new Institute of Field Physics (IOFP) at the University of North Caroline at Chapel Hill, American theoretical physicist Bryce DeWitt addressed the worldwide status of research in gravitation theory by mentioning only seven other research centers that were pursuing such activities.[1] In his account, DeWitt left out many of the research centers that were now involved in field.

This provides an indication that an emerging scientific leader did not know exactly what was happening in gravitation physics research in the various parts of the globe at that time. However, DeWitt's list says more about the dispersion of the research agendas; namely, that his perception was biased by his implicit judgment about what was to be considered relevant research in gravitation. He mentioned only centers focusing in areas he perceived close enough to—or useful to—his own field of interest: the quantization of the gravitational field. In other words, DeWitt's selection gives a good representation of how, in the mid-1950s, scientists tended to see the field as an extension of their own research pursuits.

This was precisely the kind of perception the community-building activities were most effective in modifying, as later recollections confirm. According to French mathematician André Lichnerowicz, when he entered the field in the late 1930s, "the relativity community had a strong mix. There was a small group of specialized physicists, such as W. Pauli, L. Infeld, B. Hoffman, and V. Fock; and a small group of specialized astronomers, such as G. Lemaître, and a small group of mathematicians, such as T. Levi-Civita, T. de Donder, and G. Darmois" interested in *different aspects* of the theory. "The physics community of the time, passionately involved with quantum mechanics considered relativists to be marginal" (Lichnerowicz 1992, p. 113). Lichnerowicz's view of the *relativity community* as a small, dispersed aggregate of individuals with different expertise pursuing different research agendas coincides with the perspective embraced in this book. The renaissance of general relativity was, then, for Lichnerowicz, also related to the emergence of a real *community,* but he left unexplained whether this was a consequence or a cause of the renaissance, or whether it could even be regarded as the definition itself of the renaissance process.

[1]The centers were Syracuse, Princeton, Purdue, IOFP, Cambridge, Paris, Stockholm, and RIAS (DeWitt 1957, quoted in Rickles 2011, p. 16).

Whatever Lichnerowicz's views on this, he went so far as to establish a clear demarcation about when this situation radically changed: "This state of affairs lasted until 1955 and the Bern Colloquium presided over by Pauli" (Lichnerowicz 1992, p. 113). The 1955 Bern conference, in Lichnerowicz's view, was a turning point in the renaissance process in that it modified the social structure of scholars working on topics related to the GRG domain from a set of separate research groups to a uniform community. Whereas this unambiguous demarcation can be considered to be a post hoc idealization of a single event at the end of a long, complex process of institutionalization and creation of an identifiable research domain, Lichnerowicz is right to assign a particularly relevant place in this process to the Bern conference. This event can indeed be considered to be the first step in the building of the GRG community. Yet, it was not the consequence of a coordinated plan related to the needs of the research centers working on general relativity topics. As it happened, the initial planning of the conference had nothing to do with the gradual increase in research activities related to gravitation theory in the world. Rather, the conference was a result of a combination of contextual factors, both local and international, such as the personal motivations of an isolated Swiss professor of theoretical physics, a celebratory occasion, local science policy dynamics, and changing political conditions in the international arena. What makes this process even more striking is that the Swiss theoretical physicist who initiated the process, André Mercier, was at that time not even pursuing active research in the field the conference helped (re-)launch.

4.1 The Jubilee Conference in Bern

1955 marked the fiftieth anniversary of one of the greatest achievements in theoretical physics: the formulation of the special theory of relativity by Albert Einstein. As is common knowledge, Einstein was an obscure third-level technical employee at the Patent Office in Bern when he published this milestone as one of the five papers of his *annus mirabilis* (Stachel 2005). At some point in 1953, the forty-year-old Swiss physicist André Mercier, Head of the Department of Theoretical Physics at the University of Bern, began making plans to celebrate the event with a big international conference dedicated to relativity theories in the very place where Einstein had produced this achievement.

Mercier was a complex, peculiar personality who combined his scientific pursuits with a literary, erudite sensibility and a fervent dedication to pedagogy (Mercier 1983). From his early studies, he took a genuine interest in a variety of different topics, including literature, poetry, philosophy, music, geology and, of course, physics. Eventually, he chose to specialize in the latter and earned a doctorate in physics in Geneva at the age of 22. Shortly afterwards, around the mid-1930s, he had the opportunity of an international postdoctoral education, first in Paris—where he studied with Louis de Broglie and the mathematician Élie Cartan—and then in Copenhagen with Niels Bohr. Once he returned to Switzerland, he held different positions in Zurich and Geneva up until 1939 when he was appointed Head of the Department of Theoretical Physics in Bern. Or rather, he became the Department of Theoretical Physics. For almost twenty years, he was the only member of the faculty, and the "Seminar for Theoretical Physics" consisted of only one room with a bookshelf. Since he did not have any colleagues, Mercier spent most of his time and energy teaching different courses in theoretical physics as well as attempting to make his department grow (Held et al. 1978; Held 1999). To his teaching activities, Mercier added a passionate involvement in philosophical enquiry, which he developed in a series of writings and talks as well as by joining Swiss philosophical circles.

Mercier also showed that he was willing to put a considerable amount of effort into the functioning of professional scientific societies. Since he obtained a permanent position in Switzerland, he became more and more involved in the work of the Swiss Physical Society (SPS) for which he served first as Secretary, from 1941 to 1943, then as Vice-President in 1949 to 1951, and finally as President from 1951 to 1953. In his role as President of the SPS, he was one of the national delegates in the early meetings promoted by UNESCO to launch the project of the European Council for Nuclear Research (CERN) and participated in the negotiations leading to the decision that Switzerland would host the laboratory.[2]

Mercier's institutional and teaching activities, including writing several textbooks, plus his active involvement in philosophy, did not leave much time for original research in theoretical physics. By 1953, when he began planning a conference on relativity theories, he had not produced much original theoretical work of his own. And what he did produce was not in the field of gravitation theory or general relativity. He had in fact been focusing on the problems of quantum electrodynamics, canonical formalism, and spinor calculus, on which he had become an expert since working with Cartan in the mid-1930s. The only publication addressing aspects related to general relativity was his 1940 critique of Milne's special relativistic theory of gravitation—an alternative to general relativistic cosmology that was in vogue in the 1930s, particularly in the UK (Milne 1935). Inspired by Hans Reichenbach's axiomatization of Einstein's theory of special relativity, Mercier produced an axiomatic formulation of Milne's theory, which led him to argue that general relativistic cosmology was to be preferred on the basis of epistemological considerations (Mercier 1940). It was not until after he had begun planning the conference in Bern that Mercier became slightly more active in theoretical aspects of Einstein's theory of gravitation by giving one talk on the unitary theory of gravitation and electromagnetism with a student of his (Mercier and Schaffhauser 1955). This was, however, all he did. Even at the Bern conference, he did not present a paper.

[2]"Conférence pour l'Organisation des Etudes concernant la Création d'un Laboratoire europée de Physique nucléaire, Paris, 17–21 Décembre 1951" UNESCO/NS/NUC/4, http://unesdoc.unesco.org/images/0015/001540/154028fb.pdf. Accessed 13 February 2017; and "Minutes of the First Session of the CERN European Council for Nuclear Research," Paris, 5–8 May 1952, http://cds.cern.ch/record/19494/files/CM-P00075404-e.pdf. Accessed 13 February 2017. For an accurate analysis of the first stage in the history of CERN, see Hermann et al. (1987).

Given that he was not actively involved to any great extent in research into the theory of general relativity at the time, the question remains as to what was Mercier's motivation for pursuing the project of a conference in relativity theories. Subsequently, in the late1970s, when general relativity had already become a highly respected branch of theoretical physics, Mercier would maintain that the idea of a conference was a response to his pressing questions about "what to do towards the promotion of the physics of gravitation and the theory of relativity" (Mercier 1979, p. 179). There are reasons to doubt that this explanation, which would make the conference a conscious attempt to create a kick-off event for the field, fairly represents the complexity of interconnected factors related to the planning of the conference. Ten years later, in 1988, Mercier gave a different account by stating that in 1953 he was trying to collect some ideas as a "justification" for holding such a conference (Mercier 1992, p. 109). In this second account, the main rationale for the conference seems simply to be the great opportunity presented by the anniversary, which could be exploited to organize an important scientific event in Bern. To solve this apparent contradiction and try to understand the reasons behind Mercier's unconditional and, as we will see, long-term dedication to the venture of building the GRG community, we need to delve into the multifaceted aspects of Mercier's thoughts and actions, including the evolution of his scientific, religious, and philosophical views along with his increasing passionate involvement in institutional duties in both Swiss and international contexts.

In his scientific activities, Mercier had shown that he was deeply concerned with foundational problems in quantum physics with a special emphasis on the different physical notions of time, which he also addressed in more detail in his philosophical analyses of scientific theories. Mercier believed that there was an unsolved contradiction between the notion of time in special relativity theory and the notion of time in the canonical formalism of quantum mechanics. While he attempted to address the problem from a philosophical perspective, he certainly thought that deeper discussions on the foundations of theoretical physics were needed in order for the subject to evolve (Mercier and Keberle 1949; Mercier 1950b, 1951; Wilker and Mercier 1953). This belief was closely related to his epistemological views on the methodology of scientific theorizing, and physics was to be taken as a model here (Mercier 1950a, p. 146). For Mercier, the axiomatization of a theory was the best methodological strategy to find internal contradictions and, at the same time, to develop a unifying picture. In the various theories of physics—which Mercier defines in a restrictive way as the science of the matter and its change over time—he identifies three different points of view which were difficult to reconcile: (a) the space-time viewpoint, which is employed in classical mechanics and, even more radically, in the theories of relativity; (b) the micro-energetic viewpoint, which is embodied in quantum theory in particular; and (c) the statistical viewpoint. In Mercier's epistemology of scientific method, these different perspectives were difficult to reconcile and led to different forms of axiomatization. This was a problem that, in Mercier's opinion, had to be addressed in the near future, and the link between the theory of gravitation and other areas of physics presented some severe difficulties here. There was, then, in Mercier a genuine interest in foundational questions that could have inspired him to promote a conference on relativity theories at that time. Yet this does not seem to be the only aspect of relevance in Mercier's chosen course of action.

The way the conference was organized, there were two interconnected elements that played an even greater role. The first one was related to Mercier's devotion to Christian principles and how these principles shaped his philosophy of knowledge—which encompassed epistemological, metaphysical, aesthetical, ethical, and religious considerations. The second relates to his international experience as one of the Swiss representatives in the foundational phase of CERN.

In the 1940s, following his desire to provide a unified description of the diverse ways in which a human being can gain knowledge, Mercier launched a philosophical agenda for putting science, art, and morals—intended as three different "*ways of approach to knowledge*," related, respectively, to the dichotomies true/false, beautiful/ugly, and good/bad—on the same methodological footing (Mercier 1950a, p. 27, Mercier's emphasis). In order to define a unifying methodology, Mercier argued that the essential role of mathematics within scientific theorizing provided an example that could be followed. It could lead to identifying the common foundation of science, art, and ethics in the concept of harmony, which Mercier defined, respectively, as mathematical construction, musical harmony, and religious revelation (Mercier 1950a, p. 43).

In Mercier's ecumenical worldview, science played a privileged role, however. Not only did science provide the methodological framework to interpret artistic and moral practices and ideals, but, more importantly, science and its applications also had an unparalleled impact on human life. From this state of affairs, it derived a social responsibility for science as a public good. By having the potential to improve the prosperity of the entire humanity and by its being intrinsically universalistic, science might help bring about and maintain peace among peoples (Mercier 1950a, pp. 73–76).[3] For Mercier, the social responsibility of science necessarily became the responsibility of its agents, the scientists, in all their activities in both research and education to serve as witnesses of the virtues of beauty and good. Taking his "master," Niels Bohr, as a role model, Mercier explicitly aimed at embodying the ideal figure of scientist-educator full of humanitarian sentiments and conscious of his social role in improving human life as well as in building international peace.

A few years later, he would express these views in terms of acts of love in the construction of democratic communities and, in the 1960s, Mercier declared that the freedom of human beings had to be based on moral responsibility devoted to the cause of dialogue, which was necessary to overcome human inequality. Only through the act of dialogue, and a moral disposition to this act, could knowledge and expertise flow in the same way heat flows between bodies with different temperatures as described by the law of thermodynamics (Mercier 1959, 1968).

---

[3]See also "Mercier Claims Science May Unite World Politics," *The Times Picayune, New Orleans,* 14 December 1968, HAM, folder BB 8.2. 1579.

Whereas these views were the result of a decade of elaboration and reflections on his own activity, by 1953, Mercier voluntarily assumed the responsibility of laying the foundations for a community of peers in the area of general relativity. In doing so, he built on his previous experience. His moral tension toward the responsibility of scientists had already informed Mercier's institutional activities in the early 1950s when he was involved in the international planning that led to the launching of CERN. Plausibly, Mercier perceived these discussions as the successful embodiment of his ideal of international science with its political implication in furthering peaceful cooperation between nations. Implicitly, CERN was becoming the fulfillment of the ideas he had just made public in his essay (Mercier 1950a).

This net of philosophical-scientific thinking, ethical concerns, and institutional activities forms the background behind Mercier's idea of organizing a conference on relativity theories. He did so not only for the sake of the specific scientific area, but, truth be told, to fulfill what he believed to be one of the humanitarian missions of scientists: to create an international community by favoring communication between individuals across national boundaries. From this perspective, the Bern conference was above all an act of explicit community building. But it was also somehow artificial, in the sense that it was the result of the ethical and philosophical aspirations of a single scientist who wished to build a community where there was none.

Plausibly, the post-WWII evolution of Mercier's philosophical views could have been related to the role of physicists, and physics, in the construction of weapons of mass destruction. We could speculate that he was attempting to contrast the image of military-related, destructive science with a more positive, constructive one, not linked to any form of military application. This might be the reason why, in his public talks, Mercier avoided any reference to specific events and actual political situations, preferring to stay at the most general level of a discourse in which science was considered to be a tool for unifying the world from the epistemological and even practical perspective.[4] By the same token, in the harshly divisive political context around 1950, Mercier's responsibility toward building an international scientific community could also be interpreted as a way to improve East-West relations. However, Mercier never made it explicit and, without further evidence, these remain pure speculations. By 1953, when he began plans to organize the conference, the international political situation was not so rosy, and we might assume that Mercier had at best a vague internationalist attitude without any clear plan as to how it could be implemented in practice.

All of Mercier's subsequent steps were permeated by this still undefined mixture of scientific, philosophical, and ethical motivations. In the fall of 1953, he began seeking academic and institutional support to realize his idea, aided also by his role as former President of the SPS. Uncertain about the actual scientific topics of the conference, he consulted other Swiss physicists. Their initial response was very

---

[4]"Mercier Claims Science May Unite World Politics," *The Times Picayune, New Orleans,* 14 December 1968, HAM, folder BB 8.2. 1579.

skeptical as to the need to make relativity theories the focus of an international conference. The rationale for the opposition was that special relativity no longer presented a problem and general relativity did not appear to be such a lively field worthy of a conference. Evidently, Mercier's colleagues did not share Mercier's foundational and epistemological concerns about the status of theoretical physics (Mercier 1992, pp. 109–110).

This situation changed when Mercier was able to win the support of the eminent Austrian-born theoretical physicist Wolfgang Pauli, at the time a full professor at the Eidgenössische Technische Hochschule (ETH) in Zurich. Pauli was famous for being quite a difficult person to deal with, and his open and harsh criticisms were feared by virtually every physicist who had come across him. Greatly interested in foundational problems of theoretical physics, however, Pauli embraced Mercier's idea of a jubilee conference on relativity theories in Bern and gladly accepted the position of chairman (*Präsident*) of the conference, while Mercier would serve as its secretary.

Once the decision was made, the first step was to obtain the blessing of the author of the theory the conference intended to celebrate. In November 1953, Einstein received a letter from his former country of residence containing a request for his approval of the jubilee conference as well as a warm invitation to attend. In his letter, somewhat surprisingly written in French rather than German, Mercier stressed that the organizers did not plan for the meeting to have a purely commemorative character. Rather, it was intended as a truly scientific event in which the status of the fields connected to relativity theories was to be thoroughly discussed and analyzed by invited speakers. When the letter came to the actual topic of the conference, the wording remained vague. Mercier only stressed that it would be dedicated to the domain of relativity, which "in the last half century, has been *increasing and developing in a magnificent manner*."[5] As I have shown, this was hardly the case and Mercier was painfully aware that physicists were by and large disinterested in this field. Mercier's words were only a kind attempt to glorify the works of the creator of the theories of relativity without explaining which theory of relativity he was referring to. Einstein, in fact, was left with the impression that the conference would focus on the theory of special relativity and, in his response, he stressed that due credit should also be given to Hendrik Lorentz and Henri Poincaré for their achievements.[6]

Discussing with other scientists, the organizers revealed a much more down-to-earth view of the status of the general relativity domain in contemporary physics. In a letter to Pascual Jordan, Pauli announced that the anniversary provided a good chance to collect funding to organize a conference on relativity theory and cosmology (Pauli 1999, pp. 442–443). Pauli—who had recently been awarded the Nobel Prize in Physics—seemed to recognize that, notwithstanding the general

[5]"[...] au cours de ce demi-siècle, s'est *amplifié et développé d'une manière grandiose*." Mercier to Einstein, 2 November 1953, CPAE 5-090, my emphasis. Unless otherwise indicated, all translations are mine.

[6]Einstein to Mercier, 9 November 1953, CPAE 5-092.

increase in funding for theoretical physics in the postwar period, this occasion had to be used to promote a field that was considered to be marginal compared to other branches of theoretical physics at the time. Even the aged and reclusive Einstein did not show a particularly strong interest in the project although he certainly appreciated the initiative. He promptly replied that he was glad to hear about the jubilee conference, but that his health would not allow him to be present in person.[7] After Pauli and Einstein, the third eminent figure Mercier attempted to attract was his former mentor Niels Bohr, who also provided an ethical model. Bohr was not particularly supportive either and politely declined to be involved (Mercier 1979).

However, Pauli's alliance was sufficient for Mercier to convince other ten professors of physics, mathematics, and astronomy working in Swiss universities to form an organizing committee. The committee soon began lobbying for funds, which all came from Swiss public institutions (Mercier and Kervaire 1956; Mercier 1992). Together with Pauli, Mercier began compiling a list of invited speakers and choosing the topics for the conference. The decision about who should be invited followed two different criteria: personal acquaintance of the organizers with those involved in the field and scientists' authority in a particular research area related to general relativity. A condition Pauli soon imposed was that the number of attendees had to be restricted.[8] This led to an elitist conception of the conference with a small list of keynote speakers and a slightly larger number of presenters giving shorter communications, all selected according to the two interconnected criteria mentioned above. Mercier involved the French mathematician André Lichnerowicz and theoretical physicist Marie-Antoinette Tonnelat whom he knew from his time in Paris, while Pauli kept Walter Baade and Pascual Jordan informed (Mercier 1979; Pauli 1999, pp. 442–443, 456–459). From these early contacts, the organizers began drawing up a list of keynote speakers and a related list of topics. After discussions with Pauli and the early involvement of the organizers' closest acquaintances, Mercier began making grand plans—"crazy plans" as Pauli called them[9]—in which he tried to get authoritative speakers to act as reviewers of the status of the field. When it was made public in 1954, the list of topics included (Anon. 1954):

1. Methods and general solutions of the equations of general relativity
2. Projective unitary theories and similar theories
3. Non-symmetrical unitary theories
4. Canonical formalism, general relativity and the quantization of the field
5. Mathematical structure of the Lorentz-group

---

[7]Ibid.

[8]"Pauli avait assumé cette présidence en posant la condition que la réunion ne comprendrait qu'un nombre limité de participants. Pour respecter ce vœu, il fallut renoncer a inviter certains savants qui auraient fort bien pu prendre part." A. Mercier, *Sur la Théorie de la Gravitation et de la Relativité Générale GRG,* p. 15. HAM, folder BB 8.2 1556, Dossier on GRG.

[9]"Man weiß bei Mercier nie, ob er nicht plötzlich verrückte Ideen über das Programm des Kongresses vorbringt," Pauli to Fierz, 23 April 1954 (Pauli 1999, pp. 571–572, on p. 572).

6. Cosmology:
   (a) Theoretical aspects of the questions
   (b) Experimental data on world-expansion
   (c) Secular variation of gravitation
7. Deviation of light
8. Physics and relativity

Although the list was public, the actual topics were still under debate between the organizers and the invited speakers. In a letter to Pauli, Swedish theoretical physicist Oskar Klein lamented that he did not agree to delivering what Mercier asked him to, and the final list of keynote speeches indicates that Klein was not the only one to have reservations about Mercier's plans.[10] Some of the invited speakers were in fact somewhat reluctant to work as reviewers of the status of the field, preferring to present their own personal achievements.

After the main program of keynote lectures had been defined, Mercier and Pauli began inviting other scientists to attend the conference and deliver short communications. In this phase, a third strategy of invitation was used in addition to the two selection criteria of personal acquaintance and professional authority discussed above: the organizers asked national scientific academies to send national experts of the field as their representatives at the conference. We are not sure how this came about, but in any case it was in consonance with Mercier's vision of science as a tool for furthering international collaboration and in line with his previous experience in international endeavors. The procedure closely mimicked the functioning of institutional frameworks for promoting international cooperation that were established, or re-established, in the postwar period. These were the procedural models of UNESCO, ISCU, and international unions Mercier had become familiar with through his involvement in the early phases of CERN. By sending the official invitation to the national academies rather than to individual scientists, the organizers were clearly imitating the international unions and the like, which considered the national academy of a specific country to be the recognized scientific representative of the entire country.

In some cases, the presenters of short communications were in fact selected by the national academies themselves, and the organizers had no say in this decision. This was the case, for instance, with the two Soviet representatives Vladimir A. Fock and Alexander D. Alexandrov.[11] When the organizing committee received confirmation from the Soviet Academy of Sciences that two Soviet scientists would

[10]Oskar Klein to Pauli, 6 August 1954 (Pauli 1999, pp. 739–742).

[11]Documents concerning Fock's participation in the Bern conference are in VFP, folder 174. For a more complete discussion of Fock's scientific and institutional activities in connection to the international community of scientists working on general relativity, see Jean-Philippe Martinez, Ph.D. dissertation on Vladimir Fock prepared at the University Paris 7—Paris Diderot, to be defended in 2017.

attend the event, one of whom was the well-known theoretical physicist Fock, Pauli did not even realize that Fock was "*the* famous Fock."[12]

Eventually, the format of the conference was embedded in the institutionalized forms of international cooperation that were governing the re-establishment of collaborative relations between scientists working in different countries through the mediation of national academies. Even when the speakers were not directly chosen by these academies, most of them acted as their official representatives and, consequently, as delegates of the country to which the national academy belonged. The meeting began with a long list of short, official statements from the national delegates who proclaimed the event's relevance for the progress of science and for furthering the cause of scientific internationalism.[13]

As discussed at length in Chap. 3, the organization of the jubilee conference occurred in a period of easing of East-West tensions. Consequently, the conference became one of the first international scientific events of the time that Soviet scientists were able to attend. Their presence was without doubt made possible by the role to be played by national academies envisaged by the organizers and it was an advantage that the conference was held in the neutral country of Switzerland. At around the same time as the jubilee conference, Switzerland was the host country of a series of East-West encounters. The Bern conference took place just one week before the Geneva Summit between political leaders of the Soviet Union, the United States, the United Kingdom, and France that aimed to reduce international tensions and one month before the Geneva Conference on the Peaceful Uses of Atomic Energy (Holloway 1994; Krige 2006). The organization of the Bern conference seems then to have been integrated into the process of the re-definition of the neutral role of Switzerland in both political and scientific matters in the Cold War period (Strasser 2009). While the choice of Bern as the venue of the conference was initially related only to Einstein's pioneering work in this city, Mercier was without doubt conscious of the specific political features of Switzerland in the international panorama. It is not easy to draw a clear boundary between the scientific and political aspects of the conference in the same way as it is not possible to neatly differentiate between Mercier's various motivations. The conference came to become a mixed representation of scientific and political elements. Although this might not have been the initial plan, Mercier followed the changing contexts to realize his ecumenical views.

That the conference had an intrinsic political character is confirmed by the events related to the organization of a similar celebration of the fiftieth anniversary of special relativity in Berlin. In the case of the Berlin event, the political relevance of

[12]Pauli to Christian Møller, 1 March 1955 (Pauli 1999, pp. 132–133, on p. 133, Pauli's emphasis). See also Pauli to Klein, 1 March 1955 (Pauli 1999, pp. 129–131). See also Chandrasekhara Venkata Raman to Werner Heisenberg, 9 November 1954, Nachlaß Werner Heisenberg, Rep. 93, Abteilung III, Max Planck Archiv, Berlin, folder 1704. In this letter, Raman asked Heisenberg whether he was willing to attend the conference as the representative of the Indian Academy of Sciences.

[13]"Messages from learned Societies" (Mercier and Kervaire 1956, pp. 31–37).

the planned conference was explicit from its inception as an occasion to create links between the East and West German physics communities "separated by the iron curtain."[14] Ultimately, the same political rationale for holding the conference became the reason why it was never realized. Scientists were unable to pursue this target and the plans evaporated, leading to the organization of two different but interconnected events with Max Born and Leopold Infeld as speakers in West and East Berlin, respectively (Hoffmann 1995, 1999). Evidently, as a place to promote East-West exchange, Bern functioned much better than divided Berlin, which was at the frontline of the Cold War at that precise historical moment.

The location of the event was conducive to the establishment of international links also from a linguistic perspective. Since Switzerland was a multilingual nation, organizing conferences with more than one official language was the norm. The official languages of the Bern conference on the fiftieth anniversary of special relativity were English, French, and German. With the choice of conducting the conference in these languages, the organizers were not only following a local custom but, more implicitly, were linking the conference to a prewar international tradition in which scientific exchange was governed by what historian of science Michael Gordin (2015) calls the "triumvirate" of scientific languages. This practice was becoming less common in the postwar period: English was becoming the dominant language by far, while Russian was rapidly growing—mostly in connection with the growth of the Soviet scientific community—becoming the second most used language in scientific publications. Parallel to this, the use of German and French was decreasing at the same rapid pace. The choice, natural for Swiss scientists, to maintain the "triumvirate" as the official scientific languages of the conference was in opposition to the dominance of the scientific languages employed by the two Cold War superpowers and implicitly attempted to re-establish the role of continental European countries in scientific matters. Albeit perhaps involuntarily, the three-language structure was a way to carve out a more important role for European scientists and institutions in promoting international cooperation in the climate of that particular period of the Cold War defined by a détente in East-West relations.

The construction of an image of pure international science—in implicit or explicit contrast to military-related research on nuclear weaponry—was in fact a recurrent topic in the official statements introducing the meeting.[15] The organizers and the attendees seemed to consider the general relativity domain a perfect scientific arena to explore the potential of East-West scientific communication, given its assumed irrelevance to national defense, in the same way as Switzerland could be seen as the best place to pursue these initiatives. For Fock, 1955 was the first year he was able to travel outside the Soviet zone of influence after the war. In his official statement as the delegate of the Academy of Sciences of the USSR, Fock

[14]Max v. Laue to Einstein, 16 January 1955, CPAE 16207-1 (translated in Hoffmann 1999, p. 139).

[15]See V. Moine, "Allocution de bienvenue" (Mercier and Kervaire 1956, pp. 25–26); and "Message prononcé par A.D. Fokker, Délégué de l'Académie Royale Néerlandaise" (Mercier and Kervaire 1956, pp. 35–36).

emphasized that the conference could become much more than an interesting scientific meeting; he expressed the hope that it would allow enduring scientific relations to be established.[16] In a few years, Fock's hope would be realized and he would also play a relevant role in the efforts to begin and maintain international scientific cooperation in general relativity, cosmology, and related fields.

4.2 Starting a Stable Tradition: The International Conferences on GRG

The jubilee conference in Bern was a success beyond the most optimistic expectations of the organizers. Not only did it provide an ideal venue for furthering international exchanges during the Cold War, it also allowed participants to get a clear idea of the status and dynamism of general relativity and related fields as well as to recognize its social dimension. In the words of Peter Bergmann, the Bern conference became the "clearing house for an active field of physics" (Bergmann 1956, p. 494). Thanks to the conference, many of the scientists who had established, or were on the verge of establishing, small research centers based on research programs related to general relativity began to realize that colleagues working in different research settings had similar concerns and were interested in analogous scientific questions. Although the participants were pursuing different research agendas, the meeting gave them the chance to share views and opinions, which eventually allowed some of them to understand that there were common questions related to general relativity proper that were of relevance to the different research projects (McCrea 1955; Pauli 1956; Bergmann 1956).

Lichnerowicz's report on the mathematical problems of the theory of general relativity and unified field theory, focusing on the recent advances on the Cauchy problem made by Lichnerowicz himself and his former Ph.D. student Yvonne Bruhat, was immediately regarded as the most relevant result for further advances in most of the different research agendas related to general relativity (Lichnerowicz 1956). Specific features of general relativity—including the non-linearity of Einstein's equations, their general covariant form, and the very fact that there is no background space—made it extremely difficult to solve the initial value problem, namely, to find solutions to the equations resulting from determined initial conditions. Lichnerowicz's long-term work on this problem and Bruhat's recent mathematical proof of existence and uniqueness of local solutions with respect to given initial data showed that an initial value formulation of general relativity was possible, thus opening new paths to derive exact solutions of Einstein's equation (Fourès-Bruhat 1952; Lichnerowicz 1955). Although Lichnerowicz had been

[16]"Je suis convaincu que notre réunion sera non seulement très intéressante en elle-même, mais aussi qu'elle donnera à des liens scientifiques qui seront durables." In "Message prononcé par V.A. Fock, Délégué de l'Académie des Sciences de l'URSS" (Mercier and Kervaire 1956, p. 37).

working on the problem since 1939 and Bruhat had already published her proof three years prior to the Bern event, the jubilee conference acted as a catalyst because it made it possible for the relevance of these results to be fully recognized. Pauli himself referred to Lichnerowicz's talk as the most important report ("Das Wichtigste"), particularly for the approaches it opened in quantized gravity research (Pauli 1956, p. 263).

A similar story can be told for the field of gravitational radiation research. Prior to the conference, research on this topic had come to a standstill. There was skepticism about the physical existence of gravitational waves, as some key experts doubted that such waves truly carried energy. In 1936, Einstein himself and his collaborator Nathan Rosen had also written a paper purported to demonstrate that plane gravitational waves did not exist. Although Einstein was finally convinced that there was a mistake in the paper and published a quite different result in the end, the status had not advanced much since that time (Kennefick 2005). When planning the conference, Mercier and Pauli did not even consider the topic worth of a keynote lecture. Attending the conference as the representative of the Israel Physical Society, Rosen presented only a short communication on the theme, and his conclusion was trenchant: "a physical system cannot radiate gravitational energy" (Rosen 1956, p. 175). This strong claim, made by the greatest authority in gravitational wave research of that time, could have marked the end of a research area. The Bern conference turned it instead into a new beginning. The talk about the theme and its problems in that social context made it appear to be an interesting field for further exploration. Intrigued by the discussions following Rosen's communication and by the challenges presented by the issue of the physicality of gravitational waves, mathematician Hermann Bondi decided to dedicate his new research center to this problem. He was in the process of establishing this at King's College London in close cooperation with the younger researcher Felix Pirani, who was also attending the Bern conference (see Appendix A.13.2).[17] Bondi's group would soon make important steps toward solving the issue of the existence of gravitational waves and contributed to the rapidly growing interest in this research subject (Kennefick 2007). Even in this case, then, the Bern conference acted as a catalyst for the social recognition of a shared, interesting problem worthy of further exploration.

The two specific instances of the research perspectives opened during the Bern conference reveal that the Swiss event had a fundamental function that was both social and epistemic at the same time. It was a moment of collective self-awareness in which the social perception of the epistemic status of the field was to become greatly transformed by the simple fact that different knowledge products were shared in the same space and time by a community of people who, prior to this

[17]For a discussion of these events as recollected by Bondi, see Kennefick (2007, pp. 125–126). An indirect confirmation that the chronology provided by Bondi is accurate can be found in the letter from Alfred Schild to Pirani, 24 May 1956, ASP, Box 86-27/1.

event, did not perceive themselves as belonging to any community. No radical novel results were presented at the conference. Most of them had already been published elsewhere, but they had hardly been a matter of widespread debate, while after the conference these results were rediscovered as part of a far-reaching research agenda, which was, however, still highly undefined.

Given the recognized positive outcomes of the conference, it became evident to some that the experience had to be repeated. However, there were several aspects of how the conference was organized that left some of the scholars actively working on the subject dissatisfied. The experts who could not attend were, of course, disappointed by the elitist nature of the conference, which, according to Mercier, "created injustices and [made people] jealous."[18] The younger generation of American scientists was particularly damaged by the organizers' selection. Although the United States seemed to have been the main country in this research domain in 1955, among the eighty-nine participants at the conference, nine held positions in the United States, but none of them were among the American scholars who had obtained a Ph.D. in research projects related to general relativity in the postwar period. The American attendees were mostly authoritative, established experts, and all of them but two originally came from central Europe. By the same token, the undemocratic division between keynote lecturers and presenters of short communications, combined with the bureaucratic, formal system of institutional representations was most probably considered detrimental to conducting productive scientific discussions by scholars accustomed to a different, more open and democratic academic culture, again particularly those from the United States.

Both the recognition of the importance of international conferences for the progress of knowledge in topics related to general relativity and the dissatisfaction with how the Bern conference was conducted led to the second important step in the building of the community. Only a year and half after the jubilee conference in Bern, a conference entitled "The Role of Gravitation in Physics" was organized as the inaugural event of the IOFP at Chapel Hill—a recently established institute directed by the married couple Bryce DeWitt and Cécile Morette-DeWitt. This time, although maintaining an international character, the conference had many features that clearly made it a completely different experience to the Bern conference, symbolized by the fact that Mercier—the initiator of these community-building activities—was not even invited.

First of all, as emphasized by historian of science Dean Rickles (2011, p. 19), the Chapel Hill gathering was mostly an American affair in stark contrast to the Bern conference, which was permeated by a European spirit. The American character of the conference was explicitly stressed by the organizers who proudly underlined in

[18]"Cette façon 'peu démocratique' de convoquer un congrès a l'avantage de l'intimité et assure dans une large mesure l'homogénéité d'un group d'élite. Mais elle crée des injustices et suscite des jalousies." A. Mercier, *Leçons sur la Théorie de la Gravitation et de la Relativité Générale GRG*, p. 15. HAM, folder BB 8.2. 1556, Dossier on GRG.

the first line of the program that the conference was "the first international conference of General Relativity to be held in the United States."[19] At the Chapel Hill conference, much more space was given to research pursued in the United States, mainly focusing on different approaches to the quantization of general relativity.[20] The younger generation of American theoretical physicists were able to present their work, which produced a very lively discussion among the participants. Secondly, the conference was much less formal than the Bern conference which had been inaugurated by the long sequence of national delegates' statements. Moreover, thanks in particular to the presentation by Princeton-based American physicist Robert Dicke on new possible tests of relativistic gravitational theories, the field was to be perceived as much more related to experimental endeavors in physics than it had been since its inception.[21] Even at the level of results presented, the conference was specifically based on reports of most recent works. These reports considerably enhanced the physical intuition of general relativistic predictions, notably in theoretical gravitational radiation research. Felix Pirani's presentation of the problem of measurement in general relativity provided a re-contextualization of the formula of geodesic deviation, and, consequently, its physical implications emerged. After further elaboration by Bondi and Feynman, this new perspective soon led to a broad consensus that gravitational waves do exist and carry energy, which was a very important result in the process of the renaissance of the general relativity (Pirani 2011; see also Kennefick 2007; Abbott et al. 2016).

What was also new at the Chapel Hill conference was the character of the funding bodies. In September 1956, one of Peter Bergmann's first graduate students, Joshua Goldberg, began working at the General Physics Laboratory of the Aeronautical Research Laboratories at Wright-Patterson Air Force Base in Ohio, where he started a program for funding research projects in fields related to general relativity and gravitation. As mentioned in Chap. 2, the reason for the U.S. Air Force's interest in gravity research was related to the hope that anti-gravitational devices could eventually result from this line of theoretical research. Scientists interested in general relativity were able to exploit this state of affairs to conduct their own research without promising that the military target was actually achievable (Goldberg 1992; Rickles 2011; Wilson and Kaiser 2014; Rickles 2015).[22] The Chapel Hill conference was the very first activity in this scientific domain to receive funding from the U.S. Air Force, one of the sponsors of the conference along with the Office of Ordnance Research, the U.S. Army, the National Science Foundation and IUPAP with financial support from UNESCO. As this list shows, IUPAP was

[19]"Program of the 1957 Chapel Hill Conference on The Role of Gravitation in Physics," PISGRG, Box 1.

[20]Of the thirty-seven speakers, more than two-thirds were working in American institutions (DeWitt-Morette and Rickles 2011).

[21]For a thorough discussion of Robert Dicke's contribution to the research on experimental gravity physics between 1957 and 1967, see Peebles (2017).

[22]See also Dean Rickles and Donald Salisbury, interview with Louis Witten, 17 March 2011, https://www.aip.org/history-programs/niels-bohr-library/oral-histories/36985. Accessed 12 February 2017.

the only international institution to support the conference, which was otherwise made possible by the strong involvement of American, military-related funding agencies (Goldberg 1992; Rickles 2011).

For the participants, the Chapel Hill conference meant recognition of the scientific vitality of the field, which was able to attract well-trained younger physicists as well as experts already known for their valuable contributions to other branches of theoretical physics such as John A. Wheeler—one of the greatest authorities in nuclear physics—and Richard Feynman, celebrated for his contributions to quantum field theory (DeWitt-Morette and Rickles 2011). The Chapel Hill conference was a turning point in that it allowed scientific discussions of more innovative approaches to general relativity and gravitation theory proposed by early career scholars. Yet the international character of the conference suffered because of its location and the relevance of U.S. military funding and logistical support. Many scholars from abroad were only able to come through the U.S. Air Force's logistic support provided by the Military Air Transport Service (MATS).[23] The MATS covered almost all the travel expenses of some invited scholars and enabled at least six scholars from Europe and Japan to attend. However, this organization was not able to fund scholars working in Eastern Bloc countries. At Goldberg's explicit request, his military superior replied that the Air Force "should not be directly associated with [Iron Curtain visitors] in order to avoid possible political repercussions."[24] Ultimately, no scholars from Soviet Bloc countries attended the meeting, probably for a mixture of financial and political reasons.

Despite the numerous differences between the Bern and the Chapel Hill conferences, particularly in the community of participants and their relational modalities, both conferences were extraordinary successful in their own way. Therefore, the Chapel Hill event reinforced the conviction already formed during the Swiss gathering that international conferences were necessary for the development of general relativity and related fields, and that they should be organized on a regular basis. At the end of the Chapel Hill conference, the French experts of general relativity André Lichnerowicz and Marie-Antoinette Tonnelat proposed to host the next international conference in France in two years' time. This proposal was enthusiastically accepted and, in 1959, a greater number of scientists gathered at Royaumont, near Paris. There, the full potential of these socio-scientific events was realized for the first time. The younger generation of scholars actively participated in scientific discussions and a complete international character was re-established by bringing together scientists working on both sides of the Iron Curtain. As Mercier put it, at Royaumont, "in practice all the relativists' schools of the world were represented."[25]

---

[23]Correspondence concerning the Chapel Hill conference is in CDWP, Box 4RM235. I am grateful to Dean Rickles for clarifying the content of this correspondence.

[24]Goldberg to Cécile DeWitt, 16 November 1956, CDWP, Box 4RM235.

[25]"Pratiquement toutes les écoles de relativistes du monde furent représentées." A. Mercier, *Leçons sur la Théorie de la Gravitation et de la Relativité Générale GRG*, p. 15. HAM, folder BB 8.2. 1556, Dossier on GRG.

The Royaumont conference was very long—it lasted six days—and various scientific questions were addressed from different perspectives (Lichnerowicz and Tonnelat 1962). The international character of the conference, the number of participants (around 120) and the quality of the research presented convinced the organizers of the conference and some of the senior members of the emerging community that a formal framework was needed in order to coordinate international activities and organize large conferences of this type. Enjoying the general climate of international cooperation furthered by the recent success of the International Geophysical Year, this group of senior scholars decided to follow Hermann Bondi's suggestion of creating an institutional structure to stabilize this tradition of international exchanges (Held et al. 1978). In the course of this, an imitating process similar to the one shaping the Bern conference took place. The protagonists modeled the institutional structure on the existing forms of international collaboration they knew best: the international unions and their committees on specific research fields (see Chap. 3).[26] This group of scholars established itself as the permanent International Committee on General Relativity and Gravitation (ICGRG) with the explicit task of "coordinat[ing] collaboration in scientific work in the field of General Relativity, Gravitation and related subjects, especially to help towards the organization of international Conferences and of other meetings of minor importance [throughout] the World, and to promote mutual information useful to all interested in the field" (Mercier and Schaer 1962, p. 1).

Since the inception of this institutional venture, the members employed a variety of political and scientific categories to model the ICGRG. Besides the name Committee, the members of the ICGRG also appropriated other features of the well-established committees of international scientific unions. In particular, the promoters paid special attention to the issue of national representation, by ensuring as far as possible that the different geographical areas where research in fields related to general relativity was pursued had their own representatives in the newly established body. The only relevant exceptions were Germany—both West and East—and India, with the latter not having actively taken part in the international conferences at this stage.[27]

The need to represent the different geographical areas was not the only criterion for allocating seats on the committee. The scholars who met at the international conferences were pursuing research projects related to general relativity but aimed

[26]To give a few examples of the familiarity of the ICGRG members with international structures of this type: one of the most authoritative scientists who would become member of the ICGRG, J.A. Wheeler, had been Vice-President of IUPAP from 1951 to 1954; Lichnerowicz played a role in IMU and would become President of the International Commission on Mathematical Instruction between 1963 and 1966; Tonnelat was active in the International Union on the History and Philosophy of Science and would become its Vice-President in the 1960s; Mercier was a Swiss delegate of IUPAP. Minutes of the Meeting of the ICGRG, 30 June and 7 July 1965, PISGRG, folder 1.1.

[27]The only presence of Indian scientists at the first three international conferences was Satyendra Nath Bose's institutional role as the delegate of the Indian Academy of Sciences at the Bern conference, since Heisenberg declined to attend.

at quite different targets, such as the search for unified field theories, the quantization of Einstein's equation, the development of cosmological models, mathematical issues, or questions linked to specific aspects of the theory such as gravitational radiation and exact solutions of Einstein's field equations. The ICGRG was also designed to be representative of the various areas of research within a more general, encompassing field that, from then on, was called "general relativity and gravitation" (GRG).

Scientific and geographical representativeness was only one of the explicit criteria for the composition of the ICGRG. The other was the level of authority of its members, not only within the newborn community but, most importantly, in the larger physics and mathematics communities (Mercier 1970). When the ICGRG was established in 1959, the sixteen members were: Peter Bergmann (Syracuse University), Hermann Bondi (King's College London), Carlo Cattaneo (University of Rome), Bryce DeWitt (University of North Carolina, Chapel Hill), Paul Dirac (University of Cambridge), Vladimir Fock (University of Leningrad), Leopold Infeld (University of Warsaw), Dmitri Ivanenko (University of Moscow), André Lichnerowicz (Collége de France, Paris) as Co-President, André Mercier (University of Bern) as Secretary, Christian Møller (University of Copenhagen), Nathan Rosen (Israel Institute of Technology, Haifa), Léon Rosenfeld (NORDITA, Copenhagen), John L. Synge (DIAS, Dublin), Marie-Antoinette Tonnelat (Sorbonne, Paris) as Co-President, John A. Wheeler (Princeton University). All of them met the criteria described above and were considered able to increase the prestige of the field within both national and international scientific communities.

The political character of the venture could be seen quite clearly from the fact that no scientist working on either side in the divided country of Germany was included in the list, although Pascual Jordan in Hamburg, Helmut Hönl in Freiburg, and Achilles Papapetrou in Berlin were recognized authorities in the field and, in principle, met all the criteria with which the ICGRG was being assembled (see Appendix A.3 and A.5). While it has not been possible to find any documentation concerning the formation of the ICGRG, the only rationale for not including any scholars working in Germany, particularly not Pascual Jordan, could have been political considerations, probably also related to the unsettled position of a divided Germany in a polarized Europe. From the early 1950s onward, West Germany was gradually being re-admitted to international institutions for scientific cooperation, albeit with various degrees of opposition. The Federal Republic of Germany was indeed admitted as a member of UNESCO as early as 1951 (Sewell 1975, p. 151). The East German position in international politics was much more critical as it was not officially recognized by Western nations as a separate state. The German Democratic Republic's request in 1954 to join UNESCO was in fact rejected and it would not do so until 1972, after being admitted to the United Nations (Sewell 1975, p. 181; Fennell 1994, pp. 100–101; Greenaway 1996, p. 199).

International scientific unions moved faster, but the first integration of East German scholars as representatives of their national academy was not until the early 1960s. It is possible that the members of the ICGRG wanted to avoid anything that might be construed as political implications of their choice by not inviting any scholars from either part of Germany to join them. Of course, Jordan's past sympathy for the Nazi regime and his later parliamentary activity in the Christian Democratic Union in support of nuclear armament of Germany during the third Adenauer Cabinet in 1957 did not help the German cause (see, e.g., Wise 1994; Ehlers et al. 2007; Carson 2010). Jordan's presence was certainly perceived as a threat to the attempts to develop international collaboration that was heavily dependent on the worldwide détente of East-West political tensions.

The members of the ICGRG decided that the President of the new organization would hold his or her position for three years—from one international conference to the next. The Chairman of the organizing committee of the last conference would become its President until the following conference.[28] The main pillar in this organizational scheme, however, was the Secretary of the ICGRG, André Mercier, who would hold this position for several years. As Mercier himself recognized, he was chosen principally for two reasons. First, it was a way to acknowledge the role he played as Secretary of the Bern conference in beginning the tradition of large international conferences that the ICGRG was meant to stabilize. More subtly, the decision was also made taking into consideration that Mercier was working in Switzerland, which, as Mercier emphasized, "appears as a singular domain in the surface of the Earth where things of that sort are well situated."[29] Mercier's statement reveals that he, and presumably his ICGRG colleagues, understood very well that the political neutrality of Mercier's country of residence and its established role in the diplomatic interchange with respect to the balance of power during the Cold War made Switzerland the preferred venue for scientific international ventures (Strasser 2009). Because of its neutrality and its internationally recognized role as a neutral scientific site—reinforced by the recent establishment of CERN—Switzerland was considered able to meet the needs of an international committee that also included scientists working in the Soviet Bloc countries among its members. The very existence of the ICGRG was strongly related to the developments of international political relations during the Cold War and its structure mirrored the geopolitical scenario of the time—an aspect that would become increasingly apparent in the following years.

4.3 A New Community on Paper: The Bulletin on General Relativity and Gravitation

The main explicit goal of the newly established ICGRG was to maintain and strengthen the tradition of international conferences, which were held every three years from 1959 onward. Soon, the scholars involved decided that the new

[28]The last President who was elected following this rule was Nathan Rosen in 1974, who was also the first President of the new society, the International Society on GRG (ISGRG).

[29]Mercier to Scientists throughout the World active in the field of Theories of Relativity and Gravitation, January 1961, ISGRGR.

institutional structure should do more than support the organization of these events in order to further communication between scientists interested in the field of "General Relativity, Gravitation, and related subjects" (Mercier and Schaer 1962, p. 1).

Studies on the publication rates concerning general relativity and related subjects shows that from the 1950s there was a substantial increase in the number of scientific papers published in these fields (Sect. 2.1; see also Eisenstaedt 2006, p. 248; Kaiser 2015). This was a consequence of the growing number of people working in the fields, which led to a steady rise in the number of new findings (or re-discovery of old ones).[30]

The potential expansion of the field was, however, undermined by the dispersion of the literature in publications belonging to different disciplinary domains, different national traditions, and different languages. A statistical study on the papers appearing between 1948 and 1962 revealed that around 1,500 papers had been published in over 200 journals and in six different languages, including Esperanto.[31] Moreover, the study revealed that in the 1950s the communities of scholars interested in GRG were still following publication politics related to national or even local traditions—something that was quite common at the time and not specific to this particular field. For instance, Bergmann published almost exclusively in the *Physical Review* and other journals of the American Physical Society. In turn, the *Physical Review* normally only contained papers written either by American physicists or by physicists who were working in American institutions. Lichnerowicz and Tonnelat published their papers in French journals. Infeld's preferred publication venues were Polish journals such as *Acta Physica Polonica* or the *Bulletin of the Polish Academy of Sciences*, and so on. Notwithstanding the rapid increase in publication rates, the field seemed to maintain a high degree of dispersion as far as the publication of new findings was concerned. By dispersion, here I mean that knowledge was distributed over an extremely large number of journals, some of which were accessible only to a small proportion of the community, those who were more geographically close to the source of new knowledge. To give an example, Polish theoretical physicist Andrzej Trautman—a Ph.D. student of Infeld's between 1955 and 1958—produced important findings on the topic of gravitational waves in the period 1957–1958 (Trautman 1958a, b, c). Trautman published his work in the *Bulletin of the Polish Academy of Sciences*. Despite the relevance of Trautman's findings and the fact that he wrote in English, these findings only became available to a larger community through his lectures at King's College London when he was invited there by Bondi's group.[32]

[30]Many insights gained in the previous period were re-discovered during the renaissance of general relativity. For a fitting example, see Stachel (1992).

[31]This study was done by the author on the bibliography found in the *Bulletin on GRG*.

[32]According to Trautman, his decision to publish in the venue of the Polish Academy of Sciences depended on Infeld's policy of reinforcing the prestige of this new Polish institution. Andrzej Trautman, interview with Donald Salisbury, 27 June 2016, to appear in *EPJH*. See also Felix Pirani, interview with Daniel Kennefick, 25 October 1994. I am grateful to Salisbury and Kennefick for having provided records of their interviews. An analysis of the change in the social network of scientists working in topics connected to GRG is in Lalli and Wintergrün (2016). See also Renn et al. (2016).

This dispersion of knowledge in the scientific literature was being balanced by the international conferences, smaller inter-institutional meetings, and the "postdoc cascade" process discussed in Sect. 2.3. However, within the space provided by the ICGRG meetings, some judged that these strategies were not sufficient. It was decided that this kind of dispersion of the emerging GRG field required special action. During the first meeting of the newly established committee—held in Paris in 1961—the ICGRG members envisaged a new publication venue that would serve to fill the gap in the chain of communication within the growing community. The ICGRG members decided to issue a periodical publication called the *Bulletin on General Relativity and Gravitation*, which served two different purposes: first, to publish a list of scientists working on, or interested in, the field of GRG; and second, to produce a retrospective bibliography organized through different domains of research. Mercier sent a letter to scholars on a list he had compiled—possibly at the Royaumont conference—asking them to submit their data plus a list of publications indicating the "special domain of research to which the publication refers."[33] This periodical was not a scientific journal, as it was rather analogous to news reports. The *Bulletin on GRG* only contained information useful to the scientists who were interested, but no actual research findings.

Publishing some reports with helpful information was a common strategy used by scientific institutions, including the international unions. The ICGRG members' decision to publish a bulletin seems to reveal that the main actors were again trying to follow patterns that they already considered to be established procedures for institutionalizing activities in science. However, the unsettled epistemic status of the field of GRG—namely, it being a collection of very different research agendas with no clear unified direction, pursued in research centers that had only recently started communicating with each other—led to a unique outcome. The content of the *Bulletin on GRG* was very different to what one usually found in similar publications. Typically, in bulletins, reports, and internal newsletters of international unions and their commissions, one could find descriptions of events of interest to the international community, summaries of new knowledge products, decisions concerning standards and nomenclature, and similar kinds of information. The bulletins of international scientific institutions rarely included the names and never the addresses of people working in the field, while this was the very information that the *Bulletin on GRG* was intended to circulate. The first issue was not published until Mercier had a fairly complete list of scholars in his hands. This feature made the *Bulletin on GRG* more similar to the internal organ of scientific associations than to that of an international commission based on national membership. Only in associations' reports was the list of the members of these societies sometimes published.[34]

[33]Mercier to Scientists throughout the World active in the field of Theories of Relativity and Gravitation, January 1961, ISGRGR.

[34]See, for example, *IAU News Bulletin*; the *Monthly Bulletin of Information of the ICSU;* and the section "Union News," in the journal *International Mathematical News*, which was the IMU's official News Bulletin.

Even more specific to the *Bulletin on GRG* was the list of domains of research covered under the heading of GRG as well as an indication of which specific subjects the scientists on the list were working on, which also appeared in the first issue. The list Mercier published in 1962 on the basis of the indication given by the authors included the following topics of research (Anon. 1962, pp. 3–4):

1. Canonical formalism, Lagrangians, Variation principle, etc.
2. Conservation laws, energy-momentum tensor, etc.
3. Questions of coordinates, special coordinates, etc.
4. Cosmology.
5. Electrodynamics, questions of-, Maxwell's equations. etc.
6. Experiment(s), experimental proposals or check, redshift etc.
7. Generalities, fundamentals, etc.
8. Geometrodynamics (and topology, if not under Math).
9. Linearized equations.
10. Questions of Mathematics (mathematical methods, Differential geometry, Riemannian geometry, Affine geometry, Finslerian geometry, Axiomatics etc.).
11. Hydrodynamics, Thermodynamics, Interior case, Gases, Fluids, Rigid Frames etc.
12. Equations of motion, special motions (Kepler and other).
13. N-body Problem.
14. Other theories of gravitation.
15. Particular solution(s), Classifications, Empty spaces etc.
16. Philosophy (of special or of general Relativity, Causality, Time, Space, etc.).
17. Quantization.
18. Radiation and gravitational waves.
19. Special Relativity Theory.
20. Spinors in GRG.
21. Stellar models, questions of stellar astronomy or astrophysics.
22. Unified field theory (5-dim., skew-symm., Finslerian etc.).
23. Five-dimensional unified field theory.
24. Anti-(skew-) symmetric, i.e., Einstein-Schrödinger unified field theory.

The need to create a list of the "subject matters" as well as a list of people working in the field of GRG and their sub-topics was, to the best of my knowledge, something hitherto unseen in otherwise similar publications. It was particularly characteristic of this endeavor that the classification was on the basis of what the individual authors wrote in their replies. These specific features of the *Bulletin on GRG* indicate that the ICGRG members—and Mercier as main vehicle of the ICGRG—had the conscious intent to build a community and, more implicitly, a sense of a community by creating connections between people in a field of research that was at the time both dispersed and undefined. The scientists included in the list were pursuing research agendas and the links between these different research projects were still under construction, both socially and epistemically (Blum et al. 2015). Through the organizing activity of Mercier's, the ICGRG and its publication venue became a way to define a new field called GRG, which included over twenty domains of research pursued by various research centers in different parts of a world polarized by the Cold War.

This somewhat ambitious task to counteract the social and epistemic dispersion of the field of GRG by publishing and distributing an anomalous kind of publication required the Bulletin's editors to put an enormous amount of time and effort into the project. The ICGRG's Secretary, André Mercier, was asked to administrate this activity by serving as the main editor of the *Bulletin on GRG*. From then onward, Mercier dedicated a great deal of time to activities related to the ICGRG and to the *Bulletin on GRG*, thus becoming even more central, as the key organizer, to the increasingly structured institutional framework of the GRG domain. From the first issue published in 1962 until its eventual incorporation into the new scientific journal *General Relativity and Gravitation* in 1970, the *Bulletin on GRG* became an important element in the ICGRG's community building activities, although to some of the actors its actual effects in this process did not seem to be as great as Mercier later emphasized (Mercier 1970).[35] As Secretary of the ICGRG and editor of the *Bulletin on GRG,* Mercier came to identify strongly with the work of community building. In return, he obtained a certain amount of influence on how the GRG community, and particularly its institutional representation, unfolded. Mercier's central position in the structure of community building of GRG was completely unrelated to his actual contributions to the theoretical developments of the field since he was mostly interested in providing philosophical perspectives. Mercier's centrality depended solely on his decision to offer his services to the building of the community. Thus, he was able to have an official space to pursue his subtler and more ambitious humanitarian aims which he was making public in his philosophical talks and writings: the realization of a world community through the action of dialogue and love. The *Bulletin on GRG* and the ICGRG activities became the tools that could allow him to realize his vision of a universal community of "relativists" which acted in peace toward the advancement of knowledge for the good of the entire humanity (Mercier 1959, 1968).

4.4 The Rapid Growth of the Community: New Opportunities, New Threats

In retrospect, the participants of the Royaumont conference looked at that meeting with nostalgia as a foundational event in which a new international community was firmly establishing itself. Some of the actors perceived a sense of unity between scholars who wanted to build an institutional framework to promote the field of GRG and this allowed them to overcome major differences, both scientific and political. The Royaumont conference was in fact permeated by the perception that a

[35]Personal communications by Hubert Goenner and Georg Dautcourt.

truly international platform was being built despite a divisive political context.[36] During the conference, the spirit of union was so strong that the participants also staged a performance in which they amused themselves ridiculing the abstruse features of the different approaches as well as specific traits of the different people involved.[37] This feeling of a personal, almost intimate, relationship both with the field of study and with colleagues, whatever their nationality, undoubtedly played a role in sparking the establishment of the ICGRG and certainly shaped its first few years of activity. For Mercier, the ICGRG was in fact much more than an organizing structure: it had assumed "the *spiritual* management of [the] conferences and could recommend about the works to be pursued."[38]

The elitist character of the ICGRG, so evident from Mercier's words, was also clear from the series of tasks the ICGRG members appropriated. During preparation of the following conference, the question arose as to how participation in the conferences was to be regulated. On this issue, the ICGRG members Rosenfeld and Møller, both based in Copenhagen at the time, stated that they would have preferred to restrict the conference to a small number of both invited specialists (around 100) and subjects discussed in order to increase the feeling of intimacy and free exchange of ideas. Blaming the difficulties in significantly addressing any interesting scientific question in widely attended conferences full of short presentations, Møller and Rosenfeld argued that the small concentrated format is what allows a scientific field to make important steps forward during conferences.[39] Apparently, the proposal was accepted by the other ICGRG members, but this choice created another problem. Who should then invite the speakers? Since the ICGRG was the institutional framework established in order to organize these conferences, the ICGRG members unquestionably played an important role in this decision process. Although the final list for the different conferences was ultimately negotiated with the local organizing committees, the ICGRG members had enormous power in defining which were the important lines of research to be discussed. This role was potentially a reason for conflicts between the ICGRG members, whose decision should, at least in principle, always be based on two different criteria: scientific standing and national representativeness. Both areas could have been reason for disagreement from a variety of perspectives of a political as well as scientific nature.

During the organization of the fourth international conference held in Warsaw and Jablonna in 1962, these choices did not become matter of debate, however. For

---

[36]Tonnelat to Mercier, 10 June 1972, PISGRG, folder 1.3.

[37]"Colloque de Royaumont (21–27 Juin 1959): Communications privées," PISGRG, Box 1. On the role of entertaining theatrical performances in establishing the community at the Niels Bohr Institute in Copenhagen, see Halpern (2012).

[38]"[…] assumerait désormais la direction spirituelle de tells congrès et pourrait procéder à des recommandations sur des travaux à entreprendre." A. Mercier, *Leçons sur la Théorie de la Gravitation et de la Relativité Générale GRG*, p. 15. HAM, folder BB 8.2. 1556, Dossier on GRG, emphasis mine.

[39]Rosenfeld and Møller to members of the Committee on General Relativity and Gravitation, 2 January 1961, VFP, folder 180.

instance, it was accepted that which ten scientists would come from the Soviet Union was to be decided by the Soviet Academy of Sciences.[40] The six-day conference held in Warsaw and Jablonna was again a great success with 114 participants and presentations on innovative theoretical perspectives (Infeld 1964). Many young Polish and other Eastern Bloc scholars were able to present their work to a multinational audience for the first time. There was also a strong presence of American scholars although the meeting was held in communist Poland. The U.S. Air Force even gave its support by offering Military Air Transport Service as far as Paris to many of the thirty-six American conference attendees (Goldberg 1992). The relaxed atmosphere at the conference and the strong sense of community among its participants was again presented in theatrical form with various traits of the participants openly ridiculed on stage for entertainment (Wright 2016). As for the scientific content, among the most important advances presented at the conference there were the space-time diagrams developed by British mathematician Roger Penrose, which over the next few years would become one of the major interpretative tools of general relativity (Wright 2014).

The success of the Polish meeting shows that, in the early 1960s, the ICGRG was able to further international communication and collaboration despite dramatic political events such as the Cuban Missile Crisis and the construction of the Berlin Wall. In 1968, Hermann Bondi summarized this ability to overcome political barriers as follows: "[I]n our last conferences all difficulties due to the lack of diplomatic relations between certain countries were overcome. Thus at the Jablonna conference many scientists from the German Federal Republic were present, and at the London meeting our organizing committee made it possible for several scientists from the German Democratic Republic to attend."[41] To the participants, this period might have seemed like a golden age of general relativity as far as the development of international relationships was concerned. Of course, the absence of scholars working in Germany as members of the ICGRG could have been a matter of political debate and tension. But this was not the case. At least, it did not become an issue of explicit concern. On the contrary, many scholars emphasized the high level of international collaboration, including with members of the Jordan group in Hamburg, and the relevance of this collaborative environment for producing important new findings. Ezra Ted Newman, for example, later summarized this state of affairs in very unambiguous terms: "Between 1960 and 1962 [...] the entire theory of gravitational radiation was developed by the strong interaction of many workers from Syracuse, London, Hamburg, and Warsaw via personal contacts and word of mouth communication. [...] The high quality of the science came, at least partially, from this exchange of ideas" (Newman 2005, on p. 374; see also Goldberg 2005).

This situation, which might be called the peak of the formative phase of the GRG international community, would not last long, however. One year after the GRG conference in Poland, the general research domain of GRG itself was greatly

[40]Documents in VFP, folder 184.

[41]Hermann Bondi to all members of the ICGRG, 12 July 1968, PBP.

transformed by a single scientific event, which also had a strong impact on the social and disciplinary composition of the community of scholars working on it. The rapid development of radio astronomy—mostly related to the technological advances within the framework of military radio research during World War II—was giving a new picture of the universe that was quite different from the quiet, almost immovable cosmos appearing in the range of optical astronomy. In the 1950s, it had already become clear that the universe was full of violent events (Longair 2006). In March 1963, the combination of radio and optical observations led Dutch astronomer Marteen Schmidt to announce the discovery of a new star-like object that was extremely luminous and yet very distant from the observer. The implication was that the seemingly stellar object radiated an enormous amount of energy. The newly discovered astrophysical object, the quasi-stellar radio source—soon to be christened quasar—presented an enigma for theoretical astrophysicists as to the source of the enormous amount of energy registered in the radio and optical domain. A plausible explanation for this was that the observed energy was due to the gravitational collapse of supermassive stars, which implied the theoretical necessity to describe the object by means of general relativity or a substitute gravitational theory (Hoyle and Fowler 1963).[42]

At around the same time, the leader of the newly established Center for Research in Relativity Theory at the University of Texas in Austin, mathematician Alfred Schild, was busy working on expanding GRG research activities in Texas and had succeeded in securing the establishment of a new center at the nearby Southwest Institute for Advanced Studies—the Dallas version of the famous theory-driven institute at Princeton.[43] Recommended by Schild, British mathematician and GRG expert Ivor Robinson was appointed Head of the Mathematics and Mathematical Physics Division of the Southwest Center. Robinson, together with Schild and German-American cosmologist Engelbert Schucking, also working in Austin, began making plans to celebrate the establishment of the new center with a conference in Dallas. The recently discovered astrophysical object and theoretical enquires about its properties seemed to the organizers to be the perfect, fascinating topic for the grand launch of the new research center (Schucking 2008).

Together with Peter Bergmann, the three Texas-based scientists prepared an invitation letter explicitly stating that the aim of the conference was to connect the new discovery with the developments in the GRG domain. A set of questions was circulated in which the connection between gravitational collapse, gravitational energy, the observed properties of the quasi-stellar object, and the general relativistic prediction of the inevitability of a space-time singularity of massive stellar gravitational collapse were posed in the same conceptual framework. The attempt to epistemically relate the field of high-energy astrophysics and gravitation theory had a clear social declination: the questions they wanted to address "make it imperative

[42]For accurate historical studies on the early links between astrophysical research and theoretical research on general relativity, see Israel (1987), Bonolis (2017) and references therein.

[43]"New Relativity Center," ESP, Box 3, folder University of Texas.

to bring experts from many fields together for a thorough discussion" (Robinson et al. 1965, p. v).

The desire to gather experts in general relativity theory, astronomy, and astrophysics in the same venue to discuss specific physical problems related to a single experimental object was soon characterized, by the authors of the conference themselves, as the birth of a new field: relativistic astrophysics. The title of the new research area, probably coined by Robinson, appeared in the title of the symposium that was called "Gravitational Collapse and other Topics in Relativistic Astrophysics." Held in Dallas from 16 to 18 December 1963, the event became the first in the long-standing tradition of Texas Symposia on Relativistic Astrophysics, which successfully continues to this day.[44]

Whether consciously or not, the organizers were implementing a similar form of artificial explicit community building that had characterized the organization of the Bern conference. And, also in this case, the field—and the related community, too —did not exist prior to this symposium. As with the Bern conference, the symposium in Dallas acted as the single event that precipitated the emergence of both the field and the community. Thanks to Bergmann's contribution—the only ICGRG member among the organizers—it is likely that they also made use of the experience gained through the organization of the GRG conferences and related community-building activities that had been taking place since the mid-1950s.[45] Yet the form and the goals of the event that was to spark the field of relativistic astrophysics were quite different from the GRG community-building activities. The Dallas symposium did not have any of the ideological flavors of Mercier's activities. The organization of the symposium was based purely on the logic of problem-focused and problem-solving social interactions. In addition, it was open to all participants who wished to attend, at least as listeners, while the GRG international conferences were by invitation only.

As an attempt to launch a new interdisciplinary field, the Dallas symposium was incredibly successful in immediately establishing the field as an interesting topic for established scholars as well as for younger researchers and students who entered the field just after this event. However, the community that coalesced under the heading of relativistic astrophysics was different from the GRG community that was being built in a quite different context and with different motivations. This is immediately obvious from the fact that of the sixteen ICGRG members, only five attended the Dallas symposium and one of them, Mercier, went there only for the specific purpose of writing a report for the *Bulletin on GRG* (Mercier 1964). The only other non-U.S.-based member of the ICGRG attending the symposium was the American-Israeli Nathan Rosen, with whom Robinson had worked a decade earlier.

[44]The talks given during this conference are published in two different volumes (Robinson et al. 1965; Harrison et al. 1965). For a reconstruction of the events that led to the organization of this conference, see Schucking (2008).

[45]The three Texas organizers, Robinson, Schild, and Schucking attended the Jablonna and Warsaw conference and co-authored the report on the conference for American physicists in *Physics Today* (Robinson et al. 1963).

The reasons for the big differences between the GRG and relativistic astrophysics communities depended on various factors. First of all, the physicality of the object of study made all those involved realize that the field of relativistic astrophysics was essentially a branch of physics where the role of mathematics, and even astronomy, was subsidiary, while in the GRG domain the disciplinary identity of the field was still very controversial. Second, as previously mentioned, the organizers were not interested in anything beyond the purely scientific problems themselves and the desire to increase the prestige of their own research centers. The community-building activity related to the conference was a means to achieve a scientific target and not a goal in itself, while for the GRG community this relationship was much more ambiguous, at least for some of its major institutional actors, like Mercier. Related to this, the Texas conference, albeit an international gathering, did not have the same character of internationalism that was permeating the construction of the ICGRG. The overwhelming majority of attendees were American, and not much financial aid could be provided to those coming from abroad. Eastern scholars had difficulties in attending the meeting, either for economic or political reasons. The organizers did not deal with any political impediments by attempting to settle the matter diplomatically. Rather, they openly confronted some aspects of the Soviet policies they did not approve of. Being of Jewish ancestry, Ivor Robinson was particularly sensitive to what he perceived as anti-Semitic policies implemented in the Soviet Union. As an explicit opposition to these policies, he invited three Soviet Jewish astrophysicists who were not allowed to attend conferences abroad (Vitaly Ginzburg, Iosif S. Shklovsky, and Yakov B. Zel'dovich) and insisted on sending letters to the Soviet Academy of Sciences and the Soviet consulate in Washington D.C. to put pressure on Moscow to allow them to attend the meeting. Here, the attitude of the organizers was much less diplomatic than that of Mercier and other ICGRG members. According to Schucking's recollection, this decision to invite the three Soviet astrophysicists was a way of challenging the Soviet political authorities "just to show the bastards in Moscow" in Robinson's words—miles away from the diplomacy-driven activities of the ICGRG (Schucking 2008, p. 49).

These pressures did not succeed, however. Ultimately, none of the Soviet scholars invited obtained a visa to attend the meeting, while the only Soviet present was Yakov P. Terletsky, who was already in the United States and was known more for his loyalty to the Soviet Communist party than for his skills as a theoretical physicist (Trimble 2011, p. 14).[46] Besides Terlesky, the only other Eastern Bloc scholar who attended the meeting was György Marx of the University of Budapest, making a total of only two Eastern European scholars out of 291 registered participants. This poor attendance from Soviet Bloc countries, for both political and financial reasons, was in stark contrast to the two recent international conferences of the GRG community.

[46]For Terletsky's involvement in the attempt of the Soviet Intelligence to get information about atomic energy in the aftermath of World War II, see Holloway (1996) and Aaserud (2005).

Finally, the Dallas conference became politically connoted also for a different reason that had nothing to do with the actions and expectations of the organizers. Only three weeks before the start of the conference, the President of the United States, John F. Kennedy, was assassinated in Dallas. This event shocked the entire country, and many scientists pressured the organizers to cancel the conference. The organizers decided to go ahead with the conference, and some perceived this course of action as a lack of political sensitiveness on their part and boycotted the conference.[47]

Besides all these subtle political implications related to the Texas symposium, the event was clearly perceived as a great advancement in the field of GRG and its connections to physics proper. As Austrian-born astrophysicist Thomas Gold ironically pointed out in the after-dinner speech, the possibility that the surprising new object was an entirely relativistic effect suggested that "the relativists with their sophisticated work were not only magnificent cultural ornaments but might actually be useful to science! Everyone is pleased: the relativists who feel they are being appreciated, who are suddenly experts in a field they hardly knew existed; the astrophysicists for having enlarged their domain, their empire, by the annexation of another subject—general relativity" (Gold 1965). If this was true, if the quasar was a consequence of the extreme physical implications of the theory of general relativity, it would have been, as indeed it was, the first empirical evidence of the validity of the theory besides the minor corrections to Newtonian predictions related to all the previous and contemporary tests of general relativity. From the epistemic perspective, this had far-reaching consequences as the theory could now be related to the empirical world in a completely different way with respect to the neo-Newtonian perspective dominating the previous decade. As shown elsewhere (Blum et al. 2015) and argued in Chap. 2, this was not a consequence of the discovery of quasars per se, but it was a process of preparation from both the social and epistemic perspectives that allowed scholars to draw the connections between the field of GRG—which now existed and had its practitioners—and this latest serendipitous discovery. It was not by chance that the organizing group of the Texas symposium was composed only of GRG experts without any astrophysicist.

The discovery of quasars and the rapid integration of the phenomenon within the GRG domain celebrated during the Texas symposium had, of course, an enormous potential for the development of the field and its increasing connection with physics research, which was one of the major goals of the community-building activities initiated with the Bern conference. In terms of community building, however, it also posed a great amount of tension on a structure, the ICGRG, that was being constructed on quite different bases, and these tensions were of a both epistemic and political nature. From the epistemic perspective, Gold's view that astrophysics had "annexed" general relativity as well as a direction of research focusing on astrophysics could not be fully appreciated by those who belonged to disciplinary

[47]Peter Havas to Bergmann, 29 August 1968; Havas to Bondi, 29 August 1968, PBP. See also Schucking (2008), Trimble (2011), and Virginia Trimble, personal communication.

domains other than physics, or by those who preferred to work on different research agendas. For Mercier, for example, the Texas symposia became a "supplement" to the GRG conferences, where the GRG encompassed the entire field, while relativistic astrophysics covered only a small part of it, concentrating "on a most important application at the border of physics proper and astronomy" (Mercier 1979, p. 181). From the socio-political perspective, it was evident that the community gathered in Texas was different, and these communities now had to find new ways of interacting. The disruptive impact of the rapid transformations related to the emergence of relativistic astrophysics on the activities of the GRG international community became evident during the next international conference held in London in 1965, organized by a local committee chaired by Bondi (Bondi et al. 1965).

During the London conference, which marks the transition from the formative phase to the maturity stage of the GRG community, the ICGRG met twice to discuss a number of issues. The most pressing was changes to the committee in response to requests coming from the conference participants and external funding bodies. To reflect the fundamental change related to the emergence of relativistic astrophysics, it was proposed by the organizers of the Texas symposium that an astrophysicist should be co-opted and the name of Ginzburg, who had not been able to attend the Dallas symposium, was put forward.[48]

The proposal created some friction within the ICGRG concerning the modalities for electing new members as well as the total number of members. The question was whether it was better to enlarge the ICGRG or to create a turnover with established rules for replacement of members. The Soviet members and Lichnerowicz opposed the enlargement of the committee and proposed instead to first establish new rules for the progressive change of membership. The opposing view was held by Bondi and Infeld, according to whom growing membership was a fair representation of the "worldwide intensification and development of the work in the GRG field."[49] The majority of ICGRG members agreed on this proposal and decided to expand the ICGRG to 24 members.[50] The majority evidently perceived that an inclusive strategy was the best way to maintain the ICGRG structure through the integration of the new research areas within the existing structure and established relations.

The conflict within the ICGRG members on the election of new members mirrored a certain dissatisfaction among the GRG community at large on the way in which the ICGRG had been established. Complaints began to spread that the

---

[48]The GRG conference in London was the first scientific event outside Eastern Europe that Ginzburg was allowed to attend (Khalatnikov 2012, pp. 130–132).

[49]"[…] l'intensification et le développement des travaux GRG dans le monde." Minutes of the Meeting of the ICGRG, 30 June and 7 July 1965, PISGRG, folder 1.1.

[50]The new members were Vitaly Ginzburg, Andrzej Trautman from Poland (possibly in connection with the worsening of Infeld's health status), Alfred Schild from the University of Texas at Austin as a representative of the Texas centers and relativistic astrophysics, Clive Kilmister of King's College London, and Yvonne Bruhat from Paris.

ICGRG had not been democratically elected and was therefore a self-appointed group of scholars who had not been chosen in any democratic way (Mercier 1979).

These discussions involved another delicate issue, namely, how a greater degree of institutionalization of the ICGRG should be implemented, in particular with respect to its relationships with larger international bodies that promoted scientific cooperation, and specifically with IUPAP. This had been providing funding for international conferences on GRG since the Chapel Hill conference, even before the establishment of the ICGRG. At the time of the London conference, however, IUPAP officials were requesting more formal relationships between the two institutions in order to continue this support on a firmer basis, perhaps also as a result of the increasing project-oriented policies being promoted by UNESCO (see Chap. 3). This request presented a serious problem for many of the ICGRG members because a more official affiliation to IUPAP at the institutional level would have implied identifying the field of GRG as a sub-discipline of physics at the epistemic level.

According to the mathematician Lichnerowicz, "an affiliation to the Union of Pure and Applied Physics would risk to make us solely pure and applied physicists, while the GRG research clearly regards mathematics, astronomy and mechanics as well, each of them having their own International Union."[51] This opinion was shared by Ginzburg, according to whom the field "cannot be simply defined as part of physics; GRG constitutes a larger class."[52] Following this argument that saw GRG as a field encompassing, or at least related to, many different disciplinary domains, some ICGRG members argued that IUPAP was just one of the various international unions to which their committee could adhere, along with IMU, the IAU, the International Union of Theoretical and Applied Mechanics (IUTAM) and even COSPAR. Members of the ICGRG from continental Europe who worked in disciplinary fields other than physics strongly opposed the idea that work in the field of GRG should be associated only with physics—an idea that had long been in the minds of other members of the committee, such as DeWitt and Wheeler, who were eager to see GRG recognized by their peers as an important sub-discipline of physics. Wheeler in particular had strenuously fought to show the relevance of physics to the American community, for instance, by preventing the editor of *Physical Review* from realizing his intention of prohibiting the publication of papers on too abstract topics such as general relativity and unified field theory in the mid-1950s (DeWitt-Morette 2011, p. 6). From the early 1950s, Wheeler aimed at bringing back general relativity to the field of physics (see, e.g., Misner 2010).[53]

[51]"[u]ne affiliation a l'Union de physique pure et appliquée risquerait de faire de nous des physiciens purs et appliqués seulement, alors que GRG ressortit aussi et nettement aux mathématiques, à l'astronomie et à la mécanique, qui ont chacune leur union internationale." Minutes of the Meeting of the ICGRG, 30 June and 7 July 1965, PISGRG, folder 1.1.

[52]"[...] ne peut être définie simplement comme partie de la physique; GRG constitue une classe plus vaste." Minutes of the Meeting of the ICGRG, 30 June and 7 July 1965, PISGRG, folder 1.1.

[53]In his first notebook on relativity, Wheeler proclaimed that he deemed it necessary to clearly convey the links between general relativity and "other fields of physics" to his students in preparation of his first course on general relativity. Relativity Notebook 1, JWP, Box 39 (quoted in

For this reason, he considered those who were only specialists in general relativity and did not know other aspects of physics as "'one-legged men'—men who know nothing but relativity."[54] He also harshly used to openly criticize the expression "relativists" by declaring: "There is no such thing; they are physicists" (Bartusiak 2015, p. 91). In principle, Wheeler, and along with him those who had been struggling to see the complete absorption of general relativity in the physics domain might have seen an affiliation to IUPAP as a possible strategy to establish this disciplinary affiliation at both the institutional and epistemic level, but this argument was not explicitly used to counter those expressing an opposing view. Wheeler's arguments in favor of the affiliation to IUPAP only concerned the financial and organizational support that would be generated by this affiliation.

The opposition was not just about the resistance to establishing the GRG domain as a sub-discipline of physics. Other concerns regarded the institutional changes required in order to become an official IUPAP commission, which would radically change the character of the committee by affecting the freedom allowed by the more informal character of the ICGRG. Fock complained that becoming an IUPAP commission meant that the ICGRG would turn into a more official international body. He also expressed his unease about this change. He did not explicitly explain why (or, at least, it was not reported in the minutes of the meeting), but it was evident, probably to all the participants, that his fear was that a more official structure could have made the participation of Soviet scholars more complicated from a political standpoint.

The minutes of the 1965 meeting strongly conveys the tension surrounding the different perspectives on the future of the ICGRG. A number of options were put on the table: affiliation to IUPAP, affiliation to one of the other international unions mentioned during the discussions, or the establishment of a joint commission related to more than one international union. Despite the variety of proposals, the only option that was seriously discussed as feasible was affiliation to IUPAP. Eventually, this option was also dismissed. The opposition was too strong, and its motivation so deep-rooted, that the issue could not be solved through a majority rule. In his capacity as the new President of the ICGRG, Bondi could find no better solution than postponing the decision to a future committee meeting.[55]

The recorded exchanges during the 1965 ICGRG meetings make it clear that the members of the committee were deeply concerned about how to obtain financial

(Footnote 53 continued)

Blum 2016). Later, Wheeler continued to promote the need to strengthen links between general relativity and other sub-disciplines of physics. See, for example, Wheeler to Mercier, 11 May 1967, PBP.

[54]Wheeler to Kenneth Case, 17 January 1964, JWP, Box 18, folder Misner (quoted in Bartusiak 2015, p. 91). Wheeler uses the expression "one-legged relativist" also in Wheeler to Mercier, 5 April 1961, JWP, Box 18, folder Mercier. See also Kenneth W. Ford, oral interview with John Wheeler, Session VI, 4 February 1994. Transcript available at https://www.aip.org/history-programs/niels-bohr-library/oral-histories/5908-6.

[55]Minutes of the Meeting of the ICGRG, 30 June and 7 July 1965, PISGRG, folder 1.1.

support for future conferences. These discussions, however, touched deeper themes than the purely financial factors. Some of the ICGRG members were extremely worried about the identity of GRG research. Most mathematicians in particular did not want the field to be absorbed by physics in institutional settings, pushing instead for furthering inter-disciplinary institutional frameworks, which could leave open the question as to what kind of science general relativity was. Their argument was accepted partly because there was widespread agreement that it was better not to adapt to the rules of IUPAP, which would have changed the structure, and with it the very nature, of the committee. The majority of ICGRG members only wished to secure financial support without modifying their institutional organization and losing their freedom which the more open and less rigid structure of the ICGRG offered.

In their relation to the overreaching institutional structures governing international scientific cooperation, some of the ICGRG members had contradictory views, depending on their own individual political situation. Although the ICGRG members, including the Soviets, were largely opposed to the idea of creating a more formal structure that would have changed the nature of their organization, they were nonetheless copying some of the features used in the more official international committees they did not want to resemble. In the discussions about the enlargement of the membership, it was explicitly proposed, probably by Fock, that the new members of the committee had to be elected by national delegates, thus mirroring what was one of the characteristic features of the larger international unions. The proposal was clearly a way of maintaining the decision of who should be members of this international institution in the hands of internal national organs and was something that was of particular concern to the Soviet ICGRG members. The official response to this proposal was that the ICGRG "doesn't consist of national delegates, but is constituted only by persons called to join only in virtue of their qualities as researchers and promoters of GRG."[56]

While the official policy was to maintain an independent structure made up of selected scholars chosen for their high level of expertise in GRG, in practice, unofficial policies of national representation were implemented during the 1965 meetings (Anon. 1965). The most striking evidence of attention paid to political balance in the Cold War context was that in the enlarged committee established at the London conference, there were as many Soviet members as American members —four—despite the fact that many more American scientists than Soviet ones participated in GRG activities and were on the list of GRG experts published in the *Bulletin on GRG*.[57] By implementing this balance, the ICGRG was again following

[56]"[…] ne consiste pas en délégations nationales, mais [soit] uniquement constitué par des personnalités appelées à en faire partie seulement en raison de leur qualité comme chercheurs et promoteurs en GRG." Minutes of the Meeting of the ICGRG, 30 June and 7 July 1965, PISGRG, folder 1.1.

[57]See Anon. (1962) and the corrections and/or addenda to names and addresses of scientists included in the subsequent issues of the *Bulletin on GRG* from 2 to 8. As Dieter Hoffmann (personal communication) rightly emphasized, this does not mean that there were many more

well-established Cold War rules that had been guiding the activities of international institutions—scientific or other kinds—since the Soviet Union first joined them.

Probably partly due to the pressure resulting from the discovery of quasars and the formation of a new community parallel to and only partially overlapping with the one the ICGRG believed to represent, the 1965 meeting marked a turning point for members of the ICGRG. They had to face the rapid growth and consolidation of the field of GRG on the international level as well as the establishment of relativistic astrophysics and, to do so, they had to make important decisions on key issues. Most of these decisions had subtle scientific and political implications, which led to intense disagreements between ICGRG members. These conflicts were not resolved during these meetings and remained essentially unsolved.

The last topic discussed during the ICGRG meeting was the venue of the next international conference, which would take place in 1968. Probably also as compensation for the Soviets not being able to attend the Dallas conference, the ICGRG members insisted that the next conference was in the USSR, although Fock mentioned some difficulties in organizing the conference in the Soviet Union. At the end of the second meeting, Fock promised, with the aid of his Soviet colleagues—which implied the consent of the Soviet Academy of Sciences—to ensure conditions were right for the next GRG international conference to be held in the Soviet Union in 1968. What was seen by the participants at the time as the most uncontroversial decision made during the 1965 ICGRG meetings would instead lead to events that jeopardized the success of the ICGRG's activities and would pose serious threats to its very survival.

References

Aaserud, Finn. 2005. Introduction. In *Niels Bohr—Collected works, vol. 11*, ed. Finn Aaserud, 3–83. Amsterdam: Elsevier.

Abbott, B.P., R. Abbott, T.D. Abbott, M.R. Abernathy, F. Acernese, K. Ackley, C. Adams, et al. 2016. Observation of gravitational waves from a binary black hole merger. *Physical Review Letters* 116: 061102. doi:10.1103/PhysRevLett.116.061102.

Anon. 1954. Cinquantenaire de la Théorie de la Relativité. *L'enseignement Mathématique* 40: 175.

Anon. 1962. Names and addresses of scientists. *Bulletin on General Relativity and Gravitation* 1: 3–38. doi:10.1007/BF02983128.

Anon. 1965. Communications. *Bulletin on General Relativity and Gravitation* 9: 3. doi:10.1007/BF02938032.

Bartusiak, Marcia. 2015. *Black hole: How an idea abandoned by Newtonians, hated by Einstein, and gambled on by Hawking became loved*. New Haven: Yale University Press.

(Footnote 57 continued)

American scientists working in the field than Soviet ones: since the organization of research in the Soviet Union was completely different from that in the West it was not easy to know how many scholars were working in a specific Soviet research group. For instance, the names listed in the first Soviet gravity conference held in 1961 included about 80 scientists (Garbell 1963). Of these, only six were listed in the *Bulletin on GRG* in 1962 (Anon. 1962).

Bergmann, Peter G. 1956. Fifty years of relativity. *Science* 123: 486–494.
Blum, Alexander. 2016. The conversion of John Wheeler. Talk Presented at the 7th International Conference of the European Society for the History of Science. Prague, 22 September 2016.
Blum, Alexander, Roberto Lalli, and Jürgen Renn. 2015. The reinvention of general relativity: A historiographical framework for assessing one hundred years of curved space-time. *Isis* 106: 598–620.
Bondi, Hermann, William B. Bonnor, Clive Kilmister, Ezra Newman, and Gerald Whitrow (eds.). 1965. *International Conference on Relativistic Theories of Gravitation*. London: King's College.
Bonolis, Luisa. 2017. Stellar structure and compact objects before 1940: Towards relativistic astrophysics. *The European Physical Journal H* 42: 311–393. doi:10.1140/epjh/e2017-80014-4.
Carson, Cathryn. 2010. *Heisenberg in the atomic age: Science and the public sphere*. Cambridge: Cambridge University Press.
DeWitt, Bryce. 1957. Principal directions of current research activity in the theory of gravitation. *Journal of Astronautics* 4: 23–28.
DeWitt-Morette, Cécile. 2011. *The pursuit of quantum gravity: Memoirs of Bryce DeWitt from 1946 to 2004*. Heidelberg: Springer.
DeWitt-Morette, Cécile, and Dean Rickles (eds.). 2011. *The role of gravitation in physics: Report from the 1957 Chapel Hill Conference*. Berlin: Edition Open Access.
Ehlers, Jürgen, Dieter Hoffmann, and Jürgen Renn (eds.). 2007. *Pascual Jordan (1902–1980): Mainzer Symposium zum 100. Geburtstag*. Preprint 329, MPI für Wissenschaftsgeschichte. Berlin.
Eisenstaedt, Jean. 2006. *The curious history of relativity: How Einstein's theory of gravity was lost and found again*. Princeton: Princeton University Press.
Fennell, Roger. 1994. *History of IUPAC, 1919–1987*. Oxford: Blackwell Science Ltd.
Fourès-Bruhat, Y. 1952. Théorème d'existence pour certains systèmes d'équations aux dérivées partielles non linéaires. *Acta Mathematica* 88: 141–225. doi:10.1007/BF02392131.
Garbell, Maurice A. (ed.). 1963. *Theses of the First Soviet Gravitation Conference, Held in Moscow in the Summer of 1961*. San Francisco: Garbell Research Foundation.
Gold, Thomas. 1965. Summary of after-dinner speech. In (Robinson et al. 1965), 470.
Goldberg, Joshua N. 1992. US Air Force support of general relativity 1956–72. In *Studies in the history of general relativity*, ed. Jean Eisenstaedt, and Anne J. Kox, 89–102. Boston: Birkhäuser.
Goldberg, Joshua N. 2005. Syracuse: 1949–1952. In *The universe of general relativity*, Einstein studies, vol. 11, ed. Jean Eisenstaedt, and Anne J. Kox, 357–371. Boston: Birkhäuser.
Gordin, Michael D. 2015. *Scientific Babel: How science was done before and after global English*. Chicago: University of Chicago Press.
Greenaway, Frank. 1996. *Science international: A history of the International Council of Scientific Unions*. Cambridge: Cambridge University Press.
Halpern, Paul. 2012. Quantum humor: The playful side of physics at Bohr's Institute for Theoretical Physics. *Physics in Perspective* 14: 279–299. doi:10.1007/s00016-011-0071-8.
Harrison, B. Kent, Kip S. Thorne, Masami Wakano, and John Archibald Wheeler. 1965. *Gravitation theory and gravitational collapse*. Chicago: University of Chicago Press.
Held, Alan. 1999. In memoriam: André Mercier, 1913–1999. *Foundations of Physics* 29: 1325–1326. doi:10.1023/A:1018816024789.
Held, Alan, Heinrich Leutwyler, and Peter G. Bergmann. 1978. To André Mercier on the occasion of his retirement. *General Relativity and Gravitation* 9: 759–762. doi:10.1007/BF00760862.
Hermann, A., L. Belloni, U. Mersits, D. Pestre, and J. Krige. 1987. *History of CERN, I: Volume I —Launching the European Organization for Nuclear Research*. Amsterdam: North Holland.
Hoffmann, Dieter. 1995. Die Physikalische Gesellschaft (in) der DDR. In *Festschrift 150 Jahre Deutsche Physikalische Gesellschaft*, ed. Theo Mayer-Kuckuk, 157–182. Weinheim: Wiley.
Hoffmann, Dieter. 1999. The divided centennial: The 1958 Max Planck celebration(s) in Berlin. *Osiris* 14: 138–149.

Holloway, David. 1994. *Stalin and the bomb: The Soviet Union and atomic energy, 1939–1956*. New Haven: Yale University Press.
Holloway, David. 1996. Beria, Bohr, and the question of atomic intelligence. In *Reexamining the Soviet experience: Essays in honor of Alexander Dallin*, ed. David Holloway, Norman M. Naimark, and Alexander Dallin, 235–256. Boulder, CO: Westview Press.
Hoyle, F., and W.A. Fowler. 1963. On the nature of strong radio sources. *Monthly Notices of the Royal Astronomical Society* 125: 169–176. doi:10.1093/mnras/125.2.169.
Infeld, Leopold (ed.). 1964. *Conférence internationale sur les théories relativistes de la gravitation*. Paris: Gauthier-Villars.
Israel, Werner. 1987. Dark stars: The evolution of an idea. In *Three hundred years of gravitation*, ed. Stephen Hawking, and Werner Israel, 199–276. Cambridge: Cambridge University Press.
Kaiser, David. 2015. Cold War curvature: Measuring and modeling gravitational systems in postwar American physics. Talk presented at the conference *A Century of General Relativity*, Berlin, 4 December 2015.
Kennefick, Daniel. 2005. Einstein versus the Physical Review. *Physics Today* 58: 43–48.
Kennefick, Daniel. 2007. *Traveling at the speed of thought: Einstein and the quest for gravitational waves*. Princeton: Princeton University Press.
Khalatnikov, Isaak M. 2012. *From the atomic bomb to the Landau Institute: Autobiography. Top non-secret*. Berlin: Springer.
Krige, John. 2006. Atoms for peace, scientific internationalism, and scientific intelligence. *Osiris* 21: 161–181. doi:10.1086/507140.
Lalli, Roberto, and Dirk Wintergrün. 2016. Building a scientific field in the post-WWII era: A network analysis of the renaissance of general relativity. Invited talk at the Forschungskolloquium zur Wissenschaftsgeschichte, Technische Universität, Berlin, 15 June 2016.
Lichnerowicz, André. 1955. *Théories relativistes de la gravitation et de l'électromagnétisme; relativité générale et théories unitaires*. Paris: Masson.
Lichnerowicz, André. 1956. Problèmes généraux d'intégration des equations de la relativité. In (Mercier and Kervaire 1956), 176–191.
Lichnerowicz, André. 1992. Mathematics and general relativity: A recollection. In *Studies in the history of general relativity*, ed. Jean Eisenstaedt, and Anne J. Kox, 103–108. Boston: Birkhäuser.
Lichnerowicz, André, and Marie-Antoinette Tonnelat (eds.). 1962. *Les théories relativistes de la gravitation*. Paris: Éd. du Centre national de la recherche scientifique.
Longair, Malcolm S. 2006. *The cosmic century: A history of astrophysics and cosmology*. Cambridge: Cambridge University Press.
McCrea, William H. 1955. Jubilee of relativity theory: Conference at Berne. *Nature* 176: 330. doi:10.1038/176330a0.
Mercier, André. 1940. Sur l'axiomatique de la théorie cinématique de Milne. *Helvetica Physica Acta* 13: 473–486.
Mercier, André. 1950a. *De la science à l'art et à la morale*. Paris: Neuchatel Éditions du Griffon.
Mercier, André. 1950b. Les conditions physiques et la notion de temps. *Studia Philosophica* 10: 85–114.
Mercier, André. 1951. La notion d'irréversibilité: a propos d'une analogie de la thermodynamique avec la mécanique. *Mitteilungen der Naturforschenden Gesellschaft in Bern. Neue Folge* 8: 1–27.
Mercier, André. 1959. *De l'amour et de l'etre*. Paris: Louvain.
Mercier, André. 1964. The Dallas Conference. *Bulletin on General Relativity and Gravitation* 5: 2–6. doi:10.1007/BF02938013.
Mercier, André. 1968. On the foundation of man's rights and duties. *Man and World* 1: 524–539.
Mercier, André. 1970. Editorial. *General Relativity and Gravitation* 1: 1–7. doi:10.1007/BF00759197.
Mercier, André. 1979. Birth and rôle of the GRG-organization and the cultivation of international relations among scientists in the field. In *Albert Einstein: His influence on physics, philosophy*

and politics, ed. Peter C. Aichelburg, and Roman U. Sexl, 177–188. Braunschweig: Vieweg. doi:10.1007/978-3-322-91080-6_13.

Mercier, André. 1983. *André Mercier, physicien et métaphysicien*. Berne: Institut des sciences exactes de l'Université de Berne.

Mercier, André. 1992. General relativity at the turning point of its renewal. In *Studies in the history of general relativity*, ed. Jean Eisenstaedt, and Anne J. Kox, 109–121. Boston: Birkhäuser.

Mercier, André, and Ed. Keberle. 1949. L'énergie et le temps, et le relations canoniques. *Archives des Sciences* 2: 186.

Mercier, André, and Michel Kervaire (eds.). 1956. *Fünfzig Jahre Relativitätstheorie, Verhandlungen. Cinquantenaire de la théorie de la relativité, Actes. Jubilee of relativity theory, Proceedings*. Helvetica Physica Acta. Supplementum IV. Basel: Birkhäuser.

Mercier, André, and Jonathan Schaer. 1962. General information. *Bulletin on General Relativity and Gravitation* 1: 1–2. doi:10.1007/BF02983127.

Mercier, André and Edith Schaffhauser. 1955. Théorie unitaire des champs gravitationnel et électromagnétique. In *Proceedings of the 8th international Congress of Pure and Applied Mechanics*, 521, Istanbul.

Milne, Edward Arthur. 1935. *Relativity, gravitation and world-structure*. Oxford: Clarendon Press.

Misner, Charles W. 2010. John Wheeler and the recertification of general relativity as true physics. In *General Relativity and John Archibald Wheeler*, ed. Ignazio Ciufolini, and Richard A. Matzner, 9–27. Dordrecht: Springer. doi:10.1007/978-90-481-3735-0_2.

Newman, Ezra T. 2005. A biased and personal description of GR at Syracuse University, 1951–61. In *The universe of general relativity*, ed. Jean Eisenstaedt, and Anne J. Kox, 373–383. Boston: Birkhäuser.

Pauli, Wolfgang. 1956. Schlußwort den Präsidenten der Konferenz. In (Mercier and Kervaire 1956), 261–267.

Pauli, Wolfgang. 1999. *Wissenschaftlicher Briefwechsel mit Bohr, Einstein, Heisenberg u.a. Band IV, Teil II: 1953–1954*, vol. 15, ed. Karl Meyenn. Berlin, Heidelberg: Springer.

Peebles, Phillip James Edwin. 2017. Robert Dicke and the naissance of experimental gravity physics, 1957–1967. *The European Physical Journal H* 42: 177–259. doi:10.1140/epjh/e2016-70034-0.

Pirani, Felix. 2011. Measurement of classical gravitation fields. In (DeWitt and Rickles 2011), 141–150.

Renn, Jürgen, Dirk Wintergrün, Roberto Lalli, Manfred Laubichler, and Matteo Valleriani. 2016. Netzwerke als Wissensspeicher. In *Die Zukunft der Wissensspeicher: Forschen, Sammeln und Vermitteln im 21. Jahrhundert*, vol. 7, ed. Jürgen Mittelstraß, and Ulrich Rüdiger, 35–79. München: UVK Verlagsgesellschaft Konstanz.

Rickles, Dean. 2011. The Chapel Hill Conference in context. In (DeWitt-Morette and Rickles 2011), 3–21.

Rickles, Dean. 2015. Institute of Field Physics, Inc: Private patronage and the renaissance of gravitational physics. Talk presented at the conference *A Century of General Relativity*, Berlin, 4 December 2015.

Robinson, Ivor, Alfred Schild, and Engelbert L. Schucking. 1963. Relativistic theories of gravitation. *Physics Today* 16: 17–20. doi:10.1063/1.3051062.

Robinson, Ivor, Alfred Schild, and Engelbert L. Schucking (eds.) 1965. *Quasi-stellar sources and gravitational collapse, including the proceedings of the First Texas Symposium on relativistic astrophysics*. Chicago: University of Chicago Press.

Rosen, Nathan. 1956. Gravitational waves. In (Mercier and Kervaire 1956), 171–175.

Schucking, Engelbert L. 2008. The First Texas Symposium on relativistic astrophysics. *Physics Today* 42: 46–52. doi:10.1063/1.881214.

Sewell, James Patrick. 1975. *UNESCO and world politics: Engaging in international relations*. Princeton, NJ: Princeton University Press.

Stachel, John. 1992. The Cauchy problem in general relativity: The early years. In *Studies in the history of general relativity*, ed. Jean Eisenstaedt, and Anne J. Kox, 407–418. Boston: Birkhäuser.

Stachel, John. 2005. *Einstein's miraculous year: Five papers that changed the face of physics*. Princeton, NJ: University Press Group Ltd.

Strasser, Bruno J. 2009. The coproduction of neutral science and neutral state in Cold War Europe: Switzerland and international scientific cooperation, 1951–69. *Osiris* 24: 165–187. doi:10.1086/605974.

Trautman, Andrzej. 1958a. Boundary conditions at infinity for physical theories. *Bulletin L'Académie Polonaise des Science* 6: 403–406.

Trautman, Andrzej. 1958b. Radiation and boundary conditions in the theory of gravitation. *Bulletin L'Académie Polonaise des Science* 6: 407–412.

Trautman, Andrzej. 1958c. On gravitational radiation damping. *Bulletin L'Académie Polonaise des Science* 6: 627–633.

Trimble, Virginia. 2011. The first (almost) half century of the Texas Symposia. In the 25th Texas Symposium of Relativistic Astrophysics. *AIP Conference Proceedings* 1381: 5–18. doi:10.1063/1.3635821.

Wilker, P., and A. Mercier. 1953. Remarques sur la singularité du temps, l'utilisation d'un formalisme quantique homogène et sur la relation d'incertitude entre le temps et "l'énergie". *Helvetica Physica Acta* 26: 181–190.

Wilson, Benjamin, and David Kaiser. 2014. Calculating times: Radar, ballistic missiles, and Einstein's relativity. In *Science and technology in the global Cold War*, ed. Naomi Oreskes, and John Krige, 273–316. Cambridge, MA: The MIT Press.

Wise, Norton M. 1994. Pascual Jordan: Quantum mechanics, psychology, national socialism. In *Science, technology and national socialism*, ed. Monika Renneberg, and Mark Walker, 224–254. Cambridge: Cambridge University Press.

Wright, Aaron Sidney. 2014. The advantages of bringing infinity to a finite place. Penrose diagrams as objects of intuition. *Historical Studies in the Natural Sciences* 44: 99–139. doi:10.1525/hsns.2014.44.2.99.

Wright, Aaron Sidney. 2016. Relativist nationalities: Cultures, languages, and power as performance among Cold War physicists. Talk presented at the 7th International Conference of the European Society for the History of Science, 22–24 September 2016.

Chapter 5
From Crisis to a New Institutional Body

Abstract This chapter focuses on the period between the mid-1960s and the mid-1970s, regarded as the maturity phase of the "General Relativity and Gravitation" community. During this phase, many tensions of different kinds emerged and seriously jeopardized the existence of an institutional structure for promoting general relativity at the international level. These tensions ranged from cultural differences to generational struggles, from disciplinary rivalries to political conflicts. All of them became urgent matters of debate when the international conference held in the Soviet Union in September 1968 was dramatically affected by the recent military conflicts of the Six-Day War and of the armed invasion of Czechoslovakia. Under strained political circumstances, scientists attempted to draw a clear boundary between scientific and political matters. In the attempt to do so, the participants came to hold very different views about how these demarcations should be defined in the specific context of the activities of an international scientific institution during the Cold War. Despite the various conflicts, the institution was able to survive: this period ended with the transformation of the International Committee on General Relativity and Gravitation into the International Society on General Relativity and Gravitation—whose statute came to embody the political and other tensions characterizing its establishment.

Keywords André Mercier · Anti-semitism · Christian Møller · Cold War · Czech crisis · Dmitri Ivanenko · GDR · General relativity · Hermann Bondi · International Society on General Relativity and Gravitation · Israel · Peter Bergmann · Scientific internationalism · Six-Day war · Soviet Union

5.1 From Cold-War Negotiations to Real-War Tensions: Crisis and Resolution in the Organization of the 1968 Conference in Tbilisi

Grand plans were made for the first GRG international conference held in the Soviet Union. Fock chaired the local organizing committee, which decided to host the conference at Tbilisi State University in the capital city of Georgia in early

R. Lalli, *Building the General Relativity and Gravitation Community During the Cold War*, SpringerBriefs in History of Science and Technology,
DOI 10.1007/978-3-319-54654-4_5

September 1968.[1] Many scientists working in the field of GRG awaited the Tbilisi conference with excitement for a variety of interrelated reasons. There had been major developments in the last few years that had led to enormous growth in relativistic astrophysics and observational cosmology. The serendipitous discoveries of the cosmic microwave background radiation (CMB) in 1965 and of pulsars in 1967 had provided further empirical input to link theoretical advances in gravitation and cosmology to experimental and observational activities (see, e.g., Longair 2006). Soon to be interpreted as possible empirical confirmation of the Big Bang model of the expanding universe as opposed to the Steady State theory (Dicke et al. 1965), the discovery of the CMB and its interpretation had a relevant impact in the process that led cosmology to be definitively accepted as a *physical* field of investigation. From the theoretical side, the discovery of the CMB sparked new developments aimed at explaining the observational result amidst controversy between the two cosmological models of the Steady State theory and the Big Bang theory, ultimately leading to an end to the controversy in favor of the Big Bang model around 1970 (Kragh 1996). From the experimental side, the discovery improved the testability of cosmology and gravity theories. A combination of experimental and theoretical developments quickly led to the formation of the new subfield of observational cosmology (Longair 2006; Peebles 2017). The discovery of pulsars was also soon integrated into the theoretical framework that was being built. The empirical phenomena related to these astrophysical objects led astrophysicists to immediately interpret pulsars as nucleon stars subject to enormous gravitational fields.

In theoretical astrophysics, another fundamental passage had occurred between the previous GRG conference and the Tbilisi one. Particularly in connection with a groundbreaking demonstration by Penrose (1965), consensus had been forming on the view that space-time singularities were an inescapable consequence of gravitational collapse of massive astrophysical objects. Along with the cumulative evidence supporting the existence of massive objects such as quasars and pulsars, these theoretical advances in general relativity theory led most physicists to believe in the physical existence of the entirely relativist entities described by the Schwarzschild solution. Known by the catchy name of "black holes" from 1967 onward, they soon became a major new focus of studies and provided an explanation of the phenomenon of quasars, which were then interpreted as supermassive black holes at the center of galaxies existing billions of years before they were first observed (Will

[1]Besides Fock, the local organizing committee was composed of mathematician and ICGRG member A.A. Petrov, theoretical physicist and Director of the Landau Institute for Theoretical Physics in Moscow Isaak M. Khalatnikov, the corresponding member of the Soviet Academy of Sciences, M.M. Mirianashvili, as Vice-Chairman and, as Secretary General, A.B. Kereselidze of Tbilisi State University. The reader will notice that the Soviet ICGRG member Ivanenko was not part of the local organizing committee, which is an indication of the difficult relations between Fock and Ivanenko within the Soviet group of relativists. See Jean-Philippe Martinez, Ph.D. dissertation on Vladimir Fock prepared at the University Paris 7—Paris Diderot, to be defended in 2017.

1986; Thorne 1994; Longair 2006, and references therein). All these advances let to a rapid expansion in the field of relativistic astrophysics and, more importantly for the purposes of this book, did much to consolidate the view that general relativity was a sub-discipline of physics, in consonance with what had been the clear goal of some major actors in the field, notably Wheeler (see Sect. 4.4).

The abovementioned topics were discussed prominently in the astrophysics literature and could easily be used to argue that general relativity had returned to the mainstream of physics with relevant contacts with experimental and observational activities (Wheeler 1968).[2] Wheeler had the momentum to influence the works of the GRG community, and of the ICGRG, in order to maintain a clear focus of the discussion on physics proper. He explicitly requested that the speakers invited to the Tbilisi conference should be chosen with the goal of stressing "the *connections* between this part of physics [GRG] and the other parts of physics."[3]

In terms of community-building activities, the Tbilisi conference was also considered to be an event that would demonstrate the maturity of the community created around the GRG domain. Since the conference in Poland, the existence of the ICGRG had stabilized the field from the institutional perspective. According to Mercier's recollection, it had already been decided at that point to label ICGRG-sponsored international gatherings "GR conferences" followed by the relevant conference number (Mercier 1979). A small controversy arose as to which conference should be considered to be the first: the one in Bern or the Chapel Hill conference. This indecision reflected the misgivings that some American scientists still had about the elitist, representative and formal character of the Bern conference.[4] The issue was finally resolved, probably by Mercier himself, by calling the Chapel Hill event GR1, and the Bern conference GR0 (Mercier 1979). Whatever the decision about the name and number, the Tbilisi conference was the first conference to introduce the heading "GR" in official documents and proudly print its formal title as "GR5" (Fock et al. 1968).[5] This was a clear sign that the ICGRG members and members of the local organizing committee believed the conference to be part of a tradition with an important history since the Bern conference held thirteen years previously. All these expectations for the Tbilisi conference were faced with dramatic political events that abruptly affected the world of the ICGRG, in addition to the more manageable disagreements about scientific directions and related links to particular disciplinary domains.

Unlike the symposia on relativistic astrophysics, which were open to everyone wishing to attend, participation in GRG conferences continued to be by invitation only, and the ICGRG had the task of compiling the list of participants. This was also

[2]Research on *Web of Science* shows that during the year 1968 at least 54 scientific papers contained the word "pulsar" in the title.

[3]Wheeler to Mercier, 11 May 1967, PBP, emphasis by Wheeler.

[4]Mercier, *Leçons sur la Théorie de la Gravitation et de la Relativité Générale GRG*, p. 15. HAM, folder BB 8.2, 1556, Dossier on GRG.

[5]See also "The GRn conferences," http://www.isgrg.org/pastconfs.php. Accessed 7 March 2016.

the reason why the direction of the field, as addressed through these community-building activities, depended so heavily on the negotiations within the ICGRG, and why Wheeler put pressure on its members to strengthen the focus on intra-physics links in this institutional setting.

On 22 June 1967, the ICGRG met in Paris to prepare the lists of participants and invited speakers for the Tbilisi conference.[6] During the meeting, the ICGRG members drafted a list with 128 scientists whose presence was considered "Highly Desirable" (list 2.a). They also compiled two other lists with the names of the scholars whose attendance was desirable (list 2.b) and other possible participants (list 2.c). Lists 2.a and 2.b included the names of three Israeli scholars: Asher Peres and Nathan Rosen on list 2.a and Moshe Carmeli on list 2.b, who were to receive their invitations directly from the Academy of Sciences of the USSR.[7]

In the very same month that the ICGRG met to define the ambitious project of a conference that would celebrate the successes of general relativity and, under the insistence of Wheeler and the astrophysicists, its return to true physics, a war broke out between Israel and its neighboring Arab states Egypt, Jordan, and Syria. In the six days between 5 and 10 June 1967, the third Arab-Israeli war—better known as the Six-Day War—changed the geopolitical landscape of the Middle East (Little 2010; Shlaim 2014). The dramatic situation in his own country prevented Nathan Rosen from attending the meeting in Paris. But Rosen's absence was the only effect of the third Arab-Israeli war on the ICGRG's activities at that point.[8]

However, the events that led to the war, the war itself, and its political consequences had a deep, negative impact on the relations between Israel and the Soviet Union, which in turn had dramatic repercussions on the Tbilisi conference. Despite the fact that relations between the Soviet Union and Israel had been tense since early 1953, diplomatic relations between the two countries at the ambassadorial level had continued from July 1957 to June 1967. After the Six-Day War, Soviet leaders decided to completely disrupt all diplomatic relations with Israel (Govrin 1998). The Soviet Union had been pursuing a policy of closeness with neutral Arab countries with the goal of getting them on its side in the polarized world of the Cold War, and Israel was an element of instability in the Soviet strategy. The outcome of the war, which saw the rapid and uncontroversial victory of the Israeli forces, led the Soviet Union to threaten military intervention to stop Israeli troops before the complete defeat of the Egyptian army. While the Soviet Union's possible role in the increased tension in the region that led to the outbreak of the war is still matter of historical debate, it is a fact that Israeli-Soviet relations were disrupted in the aftermath of the Six-Day War and that this was a unilateral decision made by the Soviet government (Golan 1990; Ro'i and Morozov 2008; Savranskaya and Taubman 2010; Laron 2010).

---

[6]Rosen to Mercier, 4 September 1967, PBP.

[7]Bondi to all the members of the ICGRG, 12 July 1968, PBP; copies of the lists 2.a and 2.c are preserved in PBP.

[8]Rosen to Mercier, 4 September 1967, PBP.

After the disruption of Soviet-Israeli relations, the Israeli physicist Rosen—one of the most eminent and well-respected members of the ICGRG—did something that had so far been carefully avoided by all those involved in the institutional construction of the GRG community. He explicitly addressed contingent political situations as elements that would have negative effects on the international cooperation the ICGRG was meant to promote. On 4 September 1967, he sent a letter to Mercier, copied to all the members of the ICGRG, including the Soviet members, stating that he would not attend the Tbilisi conference. Rosen explained that there was a formal problem due to the fact that he would never be able to obtain a Soviet visa because "the Soviet Union has broken off diplomatic relations with Israel."

This was only part of the problem, however. The rest of the letter contained an explicit political judgment on Soviet policies. Rosen explained that his decision also depended on questions of principle: "In view of the extreme and completely one-sided anti-Israel policy that has been adopted by the Soviet government and in view of the reports of anti-semitism as a by-product of this policy, I feel that I could not and should not visit the Soviet Union under the present conditions."[9] With his letter, Rosen explicitly put international political controversies and ethical issues on the ICGRG's agenda. Up until this point, political thinking had entered the committee's discussions only insofar as it helped to strengthen the mission of the organization to further scientific internationalism and, most notably in Mercier's case, peaceful relations. Now, the explicit criticism of Soviet domestic and foreign policies made by one of the major exponents of the ICGRG broke an unwritten rule according to which strained international political relations should be left out of the ICGRG discussions. At the time, the reaction of the other ICGRG members was unanimous: everyone rejected putting these kind of political statements on the committee's agenda by avoiding replying to Rosen's letter, at least in an official way.

In a few months, however, it would become increasingly evident that Rosen's words were both prophetic and ineffectual: he would not be invited to the conference anyway. In May 1968—less than four months before the beginning of the conference—Peres informed Peter Bergmann that none of the three Israeli scientists who had been included in the 2.a and 2.b lists had received an invitation from the USSR.[10] Although Bergmann was willing to consider the possibility that the letters of invitation had simply been lost, as had happened in other cases, he confidentially wrote to Bondi that "the Organizing Committee, either on its own initiative or at the request of the authorities in the Soviet Union, may have eliminated citizens of certain countries, and without regard to their scientific standing." In view of the fact that Bergmann considered that the ICGRG "has an obligation to see to it that our international conferences are as free of political discrimination and interference as possible," he asked Bondi, in his capacity as the President of the Committee, to

[9]Rosen to Mercier, 4 September 1967, PBP.

[10]Peres to Bergmann, 14 May 1968, PBP.

have a word with the Soviet colleagues and urge them to send the invitations to the three Israeli scientists included in the original plans.[11]

Bondi did not react promptly because he was expecting to meet Fock at a meeting in Trieste a month later.[12] Only when he discovered that he would not be going to Trieste did Bondi decide to act in an official way by sending a letter and a telegram to the Soviet members of the ICGRG urging them to forward the invitations to the three Israeli scholars on the initial lists. In his request, Bondi stressed that the vital point was "to *convince* all our colleagues that there is no discrimination against citizens of a particular country."[13] If suspicions that the lack of invitations for the Israeli scholars was politically motivated were confirmed, this would have jeopardized the entire organization of the conference.[14]

To obtain a definitive response, in early July 1968, Bondi had a telephone call with Fock.[15] At the official request of the President of the ICGRG, Fock agreed that he would ask the Academy of Sciences of the USSR to send the invitation to Peres, "though this was a little awkward because of the lack of diplomatic relations with Israel." As for the other requests, Fock refused to send an invitation to Rosen, who had already officially communicated his intention not to accept such an invitation, or to Carmeli because Fock did not regard him "sufficiently eminent." Although not happy about these last two decisions, Bondi desperately wanted to find an agreement with the Soviet colleagues as a public demonstration of the validity of the principle "that political difficulties must not stand in the way of scientific meetings." The final agreement was that the ICGRG would not withdraw its sponsorship of the Tbilisi meeting "provided, as an absolute minimum, that Peres is invited."[16] In Bondi's view, this was clearly a meaningful act that would allow the ICGRG to state that the conference held under its auspices was not excluding scientists on political grounds.

Bondi sent a report of the phone conversation to the other members of the ICGRG and asked them to confirm whether they agreed with the drastic decision to withdraw the ICGRG's sponsorship if Peres was not invited.[17] Although Bondi

---

[11]Bergmann to Bondi, 17 May 1968, PBP. See also Bergmann to Bondi, 3 May 1968, BOND, folder 4/4 A.

[12]Bondi to the members of the Committee on GRG, "The events of summer 1968," undated handwritten note, probably 4 September 1968, BOND, folder 4/4 A.

[13]Bondi to Fock, Ginzburg, Ivanenko, and Petrov, 17 June 1968, PBP, emphasis mine.

[14]Telegram from Bondi to Fock, 3 July 1968, BOND, folder 4/4A.

[15]From the documents found in the Fock Archive, science historian Jean-Philippe Martinez discovered that Fock did not reply earlier because he was not at his home institution when Bondi's communication arrived.

[16]Bondi to the Members of the ICGRG, 12 July 1968, PBP. Petrov had also replied to Bondi declaring that Rosen was not to be invited because he had officially stated that he would not have come and protested that Carmeli was not on the main list of scholars to be invited. It is very probable that Petrov was referring to the list 2.a of scholars whose presence was "Highly Desirable." Petrov to Bondi, 2 July 1968, BOND, folder 4/4A. Petrov also confirmed that they would ask the officials responsible why Peres had not received his invitation yet.

[17]Bondi to the members of the ICGRG, 12 July 1968, PBP.

professed his optimism that Soviet colleagues would invite Peres soon, it was clear that the situation was in fact a "crisis," the first serious politically related crisis in the social and institutional framework that had been created to promote the members' field of interest.[18] While Bondi was still waiting for replies from ICGRG members, he learned from American theoretical physicist Stanley Deser, who had met Fock in Trieste, that Deser had got the impression that "[Fock] neither wished nor intended to take any serious action re the Israelis. Likewise, Khalatnikov, who is running the USSR committee is in no real position to do anything without strong outside intervention since he's Jewish." Moreover, it was clear that IUPAP could not, and did not wish to, sponsor meetings where scientists of any nationality were excluded. But since "Fock and Khalatnikov alone are irrelevant," Deser thought that IUPAP needed to put direct pressure on the Soviet Academy of Sciences to solve the situation.[19]

The (few) responses to Bondi's questionnaire about his deal with the Soviets were mostly supportive, agreeing that political matters should not enter scientific meetings, although some stressed that the situation was far from ideal.[20] Bergmann was incensed by the arbitrary decision to exclude Carmeli both because it constituted a "violation of the principle of the universality of science," and because it overturned power relations between the local organizing committee and the ICGRG, which had drafted the list of scholars who should have been invited. Bergmann was willing to support Bondi's decision to negotiate this principle to "save something of the Tbilisi conference" but stressed that it would be necessary to formally change the structure of the ICGRG to avoid similar situations occurring again in the future. Bergmann requested the implementation of a formal rule according to which the ICGRG should be responsible for issuing the invitation so that the local organizing committee could not be put under pressure by its own government: "A government may still refuse to issue a visa, but it should not be permitted, as far as our influence goes, to hide behind the skirts of a scientific group which has been pressured into withholding invitations."[21]

As one might expect, Rosen was extremely disappointed with Bondi's attitude in this critical situation. He wrote to Bondi that Carmeli's exclusion was to be considered sufficient proof that political interventions were undermining the organization of the Tbilisi conference. Therefore, Rosen argued, it should not be possible to consider the Tbilisi conference as an international conference if one wanted to respect the principle that no international conference should be held in a country that did not allow certain experts to participate for political reasons. Rosen also emphasized that, in any case, the invitation to Peres might well turn out to be an "empty gesture" because Israeli scholars invited to international conferences in the

[18]The term "crisis" is explicitly used in Bondi to Member of the ICGRG, 12 July 1968, PBP.

[19]Deser to Bondi, 16 July 1968, BOND, folder 4/4A.

[20]Utiyama to Bondi, 18 July 1968; and Kilmister to Bondi, 24 July 1968, BOND, folder 4/4A.

[21]Bergmann to Bondi, 24 July 1968, PBP.

Soviet Union after the Six-Day War were usually refused their visa and could not attend anyway.[22]

In a letter to Bergmann, Rosen criticized in a more explicit and harsh way the deal between Fock and Bondi, who, according to him, was only "looking for a face-saving 'out.'" Rosen was annoyed by how people he called "liberals" changed their standards when looking at different political contexts: "It seems to me that if, for example, the Americans had been excluding Russian scientists from a Conference, there would have been a tremendous outcry instead of a readiness to settle for a symbolic single invitation."[23] As compensation for the unpleasant situation in which he and his Israeli colleagues found themselves, Rosen made the official proposal to hold the next conference in Israel.[24]

After having been reassured by Fock that an invitation would be rapidly forwarded to Peres, Bondi was so confident that he wrote an official letter to the members of the ICGRG stating that the issue would soon be resolved, although he conceded that Peres's presence would in any case remain uncertain because of the limited time remaining to solve bureaucratic problems.[25]

It turned out that Bondi was far too optimistic. On 20 August 1968, the armed troops of five countries of the Warsaw Pact—the Soviet Union, Bulgaria, the German Democratic Republic, Hungary, and Poland—invaded the Czechoslovak Socialist Republic with the aim of putting an end to the liberal reforms enacted by Alexander Dubček.[26] The invasion had such a tremendous impact around the world that the Communist parties in Western Europe either distanced themselves from the armed occupation or explicitly condemned it, a position that would shortly lead to the emergence of Eurocommunism (Schwab 1981).

In the small social world of general relativity that was being built, this aggression had dramatic consequences, too. Confusion arose as to whether the conference was to be held at all. Many people contacted Bondi's secretary to try to find out about what impact this event would have on the upcoming conference. Among them was Mercier, who wanted Bondi's authorization to forward a telegram to the ICGRG members. As Bondi was away, Mercier decided to act autonomously in his capacity of Rector of the University of Bern.[27] On 26 September, the person who

[22]Rosen to Bondi, 18 July 1968, BOND, folder 4/4A.

[23]Rosen to Bergmann, 4 August 1968, PBP.

[24]Rosen to Bergmann, 4 August 1968, PBP; and Rosen to Bondi, 8 August 1968, BOND, folder 4/4A.

[25]Bondi to ICGRG members, 6 August 1968, PBP.

[26]Although they were ready to support the operation, East German troops were actually prevented from entering the Czechoslovak national border on Soviet orders because it was feared that the memory of the German occupation during World War II would have increased Czechoslovak resistance (Stolarik 2010, pp. 137–164); see also "NVA-Truppen machen Halt an der tschechoslowakischen Grenze." http://www.radio.cz/de/rubrik/sonderserie68/nva-truppen-machen-halt-an-der-tschechoslowakischen-grenze. Accessed 17 January 2017.

[27]Browne to Bondi, 26 August 1968, BOND, folder 4/4A.

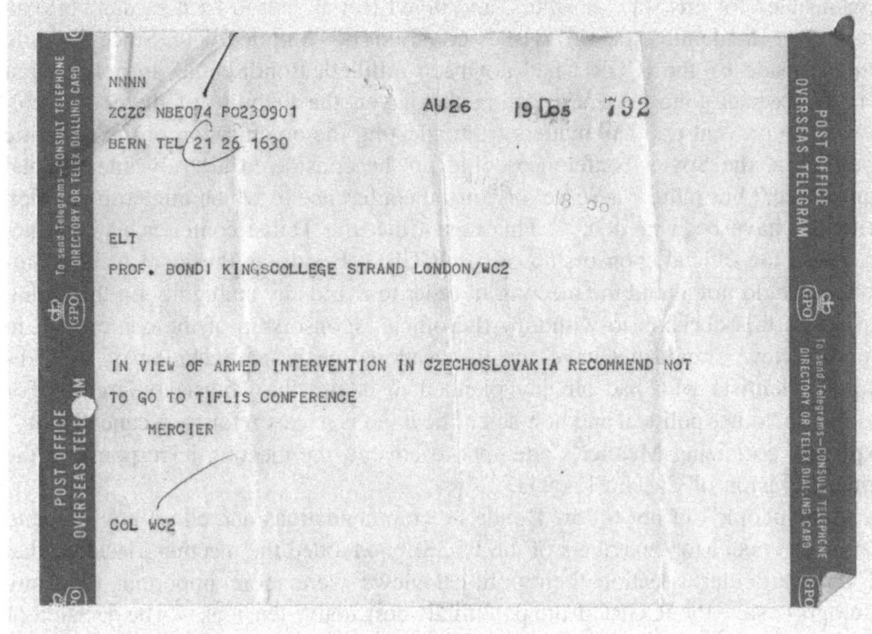

POST OFFICE OVERSEAS TELEGRAM

To send Telegrams—CONSULT TELEPHONE DIRECTORY OR TELEX DIALLING CARD

GPO

NNNN
ZCZC NBE074 P0230901 AU26 19 D05 792
BERN TEL 21 26 1630

ELT
PROF. BONDI KINGSCOLLEGE STRAND LONDON/WC2

IN VIEW OF ARMED INTERVENTION IN CZECHOSLOVAKIA RECOMMEND NOT
TO GO TO TIFLIS CONFERENCE
MERCIER

COL WC2

Fig. 5.1 Telegram Mercier to Bondi, 26 August 1968, BOND, folder 4/4A

had probably done more than anyone else to build a community through his work as Secretary of the ICGRG and editor of the *Bulletin of GRG* sent a telegram to the members of the ICGRG, stating: "[I]n view of armed intervention in Czechoslovakia recommend not to go the [Tbilisi] conference" (Fig. 5.1).[28]

This was a rather drastic step taken by a member of a formal international scientific organization who was expected to serve as a neutral secretary. Even more so as Mercier's actions in the community-building activities had always been with the conscious aim of strengthening peaceful cooperation among nations. In this specific case, however, following his personal ethical and political beliefs, Mercier felt that he had the right and the duty to profoundly affect the future of the international community by promoting a political boycott of the meeting.

To make matters worse, Peres had still not received his invitation to the GR5, which made it practically impossible to plan his participation.[29] Peres had already explicitly asked not to put the members of the Soviet organizing committee of the GR5 conference in "an embarrassing situation" because they could "not be held

[28]Telegram from Mercier to Bergmann, 26 August 1968, PBP. Mercier also sent a letter to all the scholars invited to the Tbilisi conference in which he reported the text of the telegram. Mercier to Scientist invited to partake in the Tiflis-Conference on Gravitation and the Theory of Relativity, 27 August 1968, BOND, folder 4/4A.

[29]Peres to Bondi, 20 August 1968, BOND, folder 4/4A.

responsible" for this state of affairs, and urged Bondi "not to let it escalate into an unpleasant incident."[30] Peres's wishes could not be complied with. Since the only request made by the ICGRG had not been fulfilled, Bondi could do nothing but send a six-page-long telegram stating that given the inability of the Soviet colleagues to prevent political matters from affecting the organization of the scientific conference, the Soviet conference could not be considered a "truly international conference," but rather "a Soviet-organized conference to which numerous foreign scientists have been invited."[31] This meant that the Tbilisi conference would not have had the official sponsorship of the ICGRG. Bondi, as President of the committee, would not attend the meeting in order to avoid any ambiguity on this point. However, this decision to withdraw the official sponsorship of the conference, in Bondi's view, should not have any implications on the participation of the individual scientists who had already planned to attend the Tbilisi conference. For Bondi, no "other political and non-scientific issue is a reason for such cancellation," explicitly criticizing Mercier's attempts to boycott the meeting in response to the armed invasion of Czechoslovakia.[32]

Many people did not follow Bondi's recommendations and chose not to attend the conference. Most members of the ICGRG boycotted the meeting, deciding that in this particular situation their political views were more important than any attempts to save the ICGRG from potentially destructive tensions.[33] The decision of various individuals to cancel their participation in the Tbilisi conference depended on different factors, but the Czech crisis played a particularly relevant role as a letter from Penrose to Bondi made clear: "it seemed to me that any act which appeared to condone the Russian invasion would be unthinkable."[34]

There was deep confusion and uncertainty concerning the future of the GRG community. No document can show the feelings of delusion and confusion experienced by the scholars who had long been working on strengthening the GRG field and supporting its institutional representation through the ICGRG better than an unpublished document written by Bergmann in that period. In his attempt to summarize the situation, Bergmann lamented the complete lack of coordination: everything had happened so fast that there had been no way to prepare a coordinated response to the events. Bondi indicated that he would not go because of the exclusion of the Israeli scholars; Mercier advocated boycotting the meeting to protest against the armed invasion of Czechoslovakia; the others decided individually what to do without having chance to discuss the matter amongst each other.

---

[30]Peres to Wheeler, 2 August 1968; and Peres to Bondi, 1 August 1968, PBP.

[31]Telegram from Bondi to Members of the ICGRG, 29 August 1968, PBP, also in BOND, folder 4/4A.

[32]Ibid.

[33]Peter Havas to Bondi, 29 August 1968; Goldberg to Organizing Committee, 3 September 1968; telegram from Bergmann to Fock, 3 September 1968, PBP.

[34]Penrose to Bondi, 28 August 1968, BOND, folder 4/4A. A similar motivation is to be found also in the telegram from Jules Géhéniau to Bondi, 30 August 1968; and telegram from Alfred Schild to Bondi, 30 August 1968, BOND, folder 4/4A.

This strained situation and the impossibility to organize a coordinated response to the events were putting a tremendous amount of stress on the ICGRG. It was very much up in the air whether this organization could continue in the future to promote international cooperation as it had done before these dramatic political events.[35]

Probably in response to Bondi's telegram withdrawing sponsorship by the ICGRG, which also meant a lack of financial support for the organization and some of the participants, five days before the opening of the conference, the long-awaited letter of invitation from the Soviet Academy of Sciences reached Peres.[36] Bondi soon seized the opportunity to declare the crisis resolved: the Tbilisi conference could now be officially considered as sponsored by the ICGRG and it was proposed to elect the chairman of the conference, Fock, as the future ICGRG President (Fig. 5.2).[37] It did not matter that Peres could not go because it was not feasible to receive a visa in time.[38] It did not matter that many people had already changed their plans and decided not to go, among them many members of the committee, including Bondi himself.[39] As President of the ICGRG, Bondi was relieved that he could officially save the international relations within the scientific world of the GRG community. At least formally, what Bondi had defined as political interferences in the conference organization had been successfully overcome, which allowed him to declare the conference free, in principle, from inadmissible political intervention.

Ultimately, the Tbilisi conference turned out to be an important event despite the large number of American and European scholars who decided not to go. Amongst those who did attend was a group of American physicists most closely related to Wheeler's group.[40] For the scientific discussion between this American group and

[35]Bergmann, draft, 2 September 1968, PBP; as far as I know, the document has not been used or circulated.

[36]Telegram from Peres to Bondi, 5 September 1968; telegram from Kereselidze to Bondi, 4 September 1968; Miss Speathe to Bondi, 4 September 1968, BOND, folder 4/4A.

[37]Telegram from Bondi to Bergmann, 6 September 1968, PBP. Various telegrams to ICGRG members, members of the Organizational Committee of the Tbilisi conference and to the Secretary of the Soviet Academy of Sciences, 6 September 1968, BOND, folder 4/4A.

[38]Rosen to Fock, 15 October 1968, PBP.

[39]Bondi was unable to attend the meeting because he had, in the meantime, taken other commitments as the newly elected Director General of the European Space Research Organization (ESRO).

[40]"As a whole, the Conference in Tbilisi was very pleasant and successful, although, of course, we were missing a large number of colleagues from Europe and U.S.A. It was very unfortunate that the politics of the Great Powers were able to interfere with the unity of scientists, which has worked so well in our field of research since the Berne Conference in 1955." Møller to Bondi, 15 October 1968, CMP, Box A-D, folder 3. See also Fock to Bondi, 11 November 1968, BOND, folder 4/4A. In addition to Wheeler, among the attendees were Dieter Brill, John Bardeen, Arthur Komar, Bruce Partridge, Abe Taub, Bryce DeWitt, Frederik Belinfante, Remo Ruffini. See Ruffini (2010) and Georg Dautcourt, "Bericht über die fünfte international Gravitationskonferenz in Tbilisi vom 9.-13. September 1968," DAUT.

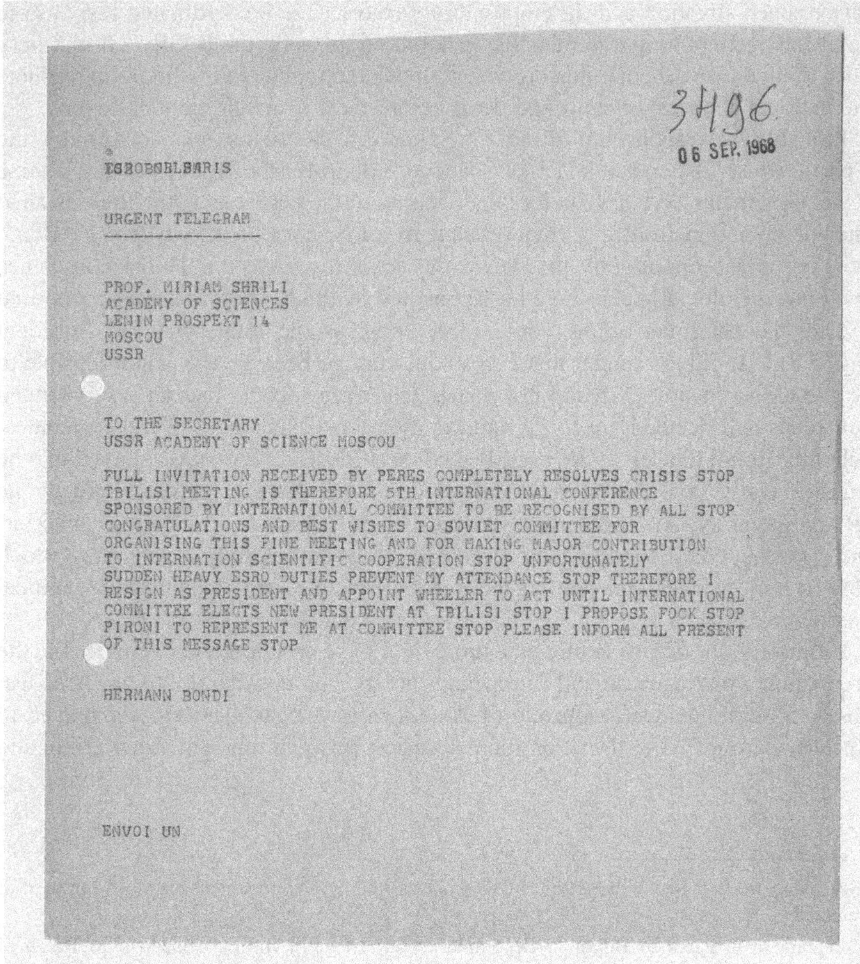

3496
06 SEP. 1968

ESROBGBLBARIS

URGENT TELEGRAM

PROF. MIRIAM SHRILI
ACADEMY OF SCIENCES
LENIN PROSPEKT 14
MOSCOU
USSR

TO THE SECRETARY
USSR ACADEMY OF SCIENCE MOSCOU

FULL INVITATION RECEIVED BY PERES COMPLETELY RESOLVES CRISIS STOP TBILISI MEETING IS THEREFORE 5TH INTERNATIONAL CONFERENCE SPONSORED BY INTERNATIONAL COMMITTEE TO BE RECOGNISED BY ALL STOP CONGRATULATIONS AND BEST WISHES TO SOVIET COMMITTEE FOR ORGANISING THIS FINE MEETING AND FOR MAKING MAJOR CONTRIBUTION TO INTERNATION SCIENTIFIC COOPERATION STOP UNFORTUNATELY SUDDEN HEAVY ESRO DUTIES PREVENT MY ATTENDANCE STOP THEREFORE I RESIGN AS PRESIDENT AND APPOINT WHEELER TO ACT UNTIL INTERNATIONAL COMMITTEE ELECTS NEW PRESIDENT AT TBILISI STOP I PROPOSE FOCK STOP PIRONI TO REPRESENT ME AT COMMITTEE STOP PLEASE INFORM ALL PRESENT OF THIS MESSAGE STOP

HERMANN BONDI

ENVOI UN

Fig. 5.2 Telegram Bondi to M.M. Mirianashvili, 5 September 1968, BOND, folder 4/4A

the Soviet scholars particularly interested in astrophysics and experimental endeavors, the meeting was very productive and would even lead to long-lasting cooperation.[41] As far as building an international community and the activities of

---

[41]The most important of Soviet-American collaborations would probably become the cooperation between American theoretician Kip Thorne and Soviet experimental physicist Vladimir B. Braginsky who had started planning gravitational wave experiments back in the early 1960s. Kip Thorne, whose leftist parents had visited the Soviet Union in the past, paid a visit to the Braginsky group in conjunction with the Tbilisi conference. In the following years, Thorne continued to visit Braginsky and his group, and they also began co-authoring papers in the mid-1970s; see Kip S. Thorne, "Vladimir Borisovich Braginsky 1931–2016," https://www.ligo.caltech.edu/news/ligo20160426#Remiscences%20by%20Kip%20Thorne. Accessed 9 March 2017. According to

the ICGRG are concerned, however, the meeting was the lowest point since the formation of the institutional body. Of the 19 ICGRG members from non-socialist states, only three attended the Tbilisi meeting: DeWitt, Møller, and Wheeler. Together with the four Soviet members and Trautman from Poland, they held a very reduced version of the ICGRG meeting, with the decisions taken considered to be provisional and subject to change if the majority of members opposed them.

Despite the small number of attendees, the dramatic events preceding the conference produced a series of resolutions that, if implemented, would have far-reaching consequences on the regulations governing the ICGRG and subsequent international conferences. First, a motion was approved that drastically expanded the participation in future GR conferences by determining that "[e]ach International Conference is open to everyone."[42] This had been a matter of contention since the establishment of the ICGRG, when the proposal signed by Møller and Rosenfeld to limit participation in the conferences to a fixed number of invited scholars had been accepted. The changing social dimension of the GRG community and the recent dramatic events indicated a need to modify one of the rules that most strongly characterized the GRG international meetings. A new formal rule was included, stating that the ICGRG was to make the final decision about the list of invited scholars, but that a scientist could solicit an invitation from the ICGRG if he or she had not already been invited. Second, it was accepted by all the attending members that "sponsorship of the conference by the International Committee implies that the host country makes timely entrance possible for every scientist recommended for participation by the International Committee."[43] This motion was the most direct response to the Israeli affair. Everyone was aware that the internal rules of an international scientific organization could not really have an impact on the policies of the hosting countries as to whether they provided visas or refused them on political grounds. Nonetheless, the new formal regulation was intended as a means of institutionalizing the fact that there was an agreement between the different members of the ICGRG, including the Soviet ones, that every possible step would be taken to avoid a repeat of what had happened in the run-up to the Tbilisi conference in the future.

The participating members also attempted to resolve the issue of replacing members of the ICGRG, which had long been considered a controversial matter. At the London conference, Mercier had proposed to solve this issue, again taking into consideration the East-West political balance, by introducing a rule on changes in membership of countries with more than three members, namely, the United States

(Footnote 41 continued)

Thorne's recollection, the relationship with Braginsky was instrumental in turning his interest toward the problem of gravitational waves, and consequently to its entrance into the LIGO venture. Another relevant relationship that started in Tbilisi was that between Thorne and Yakov B. Zel'dovich; see also Thorne to Wheeler and Charles Misner, 22 September 1969, JWP, Box 18, folder Misner.

[42]Minutes of the ICGRG during the GR5 conference held in Tbilisi on 12 September 1968, PBP.

[43]Ibid.

and the Soviet Union. Mercier's proposed rule stated that for these countries, one of the members had to retire at the end of a three-year period. The retiring member would then be replaced by another member of the same country chosen by the ICGRG members of that country only. Mercier's proposal was clearly in line with the Soviet scholars' wishes, taking into account the constraints resulting from Soviet internal policies concerning their participation in international ventures. The issue had already arisen three years earlier when the ICGRG members' proposal to elect Ginzburg as a member of the committee was made (see Sect. 4.4).[44] The procedure adopted during the election of Ginzburg in 1965—namely, the democratic selection of Soviet members by the members of the ICGRG at large—was in open opposition to the Soviet policies concerning the official participation of USSR scientists in international organizations. The structure of Soviet society and scientific research, and the central regulations concerning international relations meant that the decision about who were to be the Soviet members of international structures had to be made internally, by Soviet organs alone, in the same way the USSR participated in other international scientific institutions, such as ICSU and the international unions, namely, within the framework of national delegations.[45]

The meeting ended with the election of the new President and the decision about the venue for the next conference. Despite the controversial matters preceding the Tbilisi conference, Fock was confirmed as President of the ICGRG following the tradition that the chairman of the local organizing committee would become the committee's President until the next conference. As for the venue of the next conference, the Soviets opposed Rosen's proposal to organize it in Haifa, and those who attended the meeting agreed that it should be in a "sufficiently neutral place." Therefore, Møller was asked to look into the possibility of organizing it in Copenhagen.[46]

The events related to the Tbilisi conference were clearly leading to a change in the nature of the ICGRG and of the related community-building activities by introducing explicit negotiations about political matters in the accepted discourse among members. Since the Bern conference, in the process of developing the ICGRG and its activities, politics had always been part of how some participants saw their role in the institutional body governing a new international community. Up until 1967, however, politics had entered the discussion only implicitly and mostly uncontroversially as the background contest in which the participants had to act. In this context, there was a tacit consent to avoid any explicit debate on political matters and an attempt to, again implicitly, understand the other countries' limitations and constraints.

From the moment Rosen introduced a criticism of Soviet anti-Semitic policies into the committee's official documents, a work began, first covertly, and then

[44]Minutes of the Meeting of the ICGRG, 30 June and 7 July 1965, PISGRG, folder 1.1.

[45]Minutes of the ICGRG during the GR5 conference held in Tbilisi on 12 September 1968, PBP.

[46]Møller to Bondi, 15 October 1968, CMP, Box A-D, folder 3.

openly, to demarcate the boundary between inadmissible political intervention and what was instead acceptable politically motivated behavior in the activities of an international scientific institution acting in a conflicted political world.[47] The individual reaction to the 1967/1968 political events was very different for different actors, which also implied that the boundary-work had to become more and more explicit.

The moment the explicit definition of a demarcation of boundaries became a necessary part of the community building was related to Mercier's decision to send a telegram requesting a general boycott of the meeting as a response to the Czech crisis. The Secretary's decision clashed with the views of the President, Bondi, who was instead trying to "save" the conference despite what he considered to be a clear, insurmountable political intervention: not admitting scholars for political reasons.[48] In the dramatic days leading up to the conference and the weeks afterwards, Mercier, Bondi, and others had the chance to share their views again in order to better define the rationale behind their actions as well as their views about the demarcation between science, politics, and ethics both at the level of individual thinking and institutional representations.

In his letter to the new President, Fock, Mercier stated that he acted as he did not only in order to protest against the armed invasion of Czechoslovakia, but also because he was convinced that it would not have been possible to openly discuss the events during the Tbilisi conference. He considered his drastic calling for a boycott a way of forcing discussion between his colleagues in the Soviet Union. This action, Mercier believed, should have been even more powerful in view of the fact that everyone knew his strong commitment to friendship between peoples, between the members of the GRG family at large, and the ICGRG members in particular. The calling for a boycott was, in his view, a way to affirm that "it is wrong to aim at complete separation of science and ethics as had been done at the beginning of this century by scientists who did not understand the dangers of Nazism."[49]

Where should one then draw the boundary between science and morals in a period of political, ideological, and, often, armed conflict? How can one clearly define what is moral in politics? Why should one protest after the invasion of Czechoslovakia and not in response to the American actions in the Vietnam War? These are the questions Fock asked in response to Mercier's letter. Since it was not possible to answer these questions in any straightforward, objective manner, Fock promised to act according the principle: "*science has to be posed above politics*," and that the work of the ICGRG during his chairmanship would be based on such a principle. In this sense, Bondi's request to invite Israeli scholars, whatever position the hosting country might have about Israeli policies in the Middle East, was in

[47] At the metaphorical level, this kind of boundary-work can be assimilated to the boundary-work discussed in Gieryn (1999).

[48] The term "save" in this context was used in Bergmann to Bondi, 24 July 1968, PBP.

[49] Translation of the letter from Mercier to Fock, 3 October 1968, BOND, folder 4/4A.

Fock's view absolutely legitimate: it was the same principle—science above politics—applied to the pursuit of international scientific relations.[50]

Bondi completely agreed with the views about science and politics in matters of international relations exposed by Fock.[51] Bondi himself had already attempted to clarify his views to the ICGRG members in a long letter in which he elucidated the rationale behind his actions and his strong disagreement with Mercier. The letter—entitled "the events of summer 1968" and written before the conference started—is a fine attempt of boundary-work in the middle of a tense situation. The only real principle one had to follow was that "an essential condition for any internationally sponsored meeting [was] that there must be no exclusion of scientists on grounds of nationality, creed, or race." This was a principle that had already been institutionalized at the level of international unions. It was the only rule, Bondi believed, one should follow in deciding whether countries were to be excluded from hosting international meetings under the sponsorship of an international scientific body like the ICGRG. This was the principle he followed in that strained period and for which he struggled by urging the immediate invitation of Peres against the evident resistance of Soviet authorities. For Bondi, it was the responsibility of the organizing committee to arrange with the host government that this principle was followed. But there had always been problems here in almost every country in different periods during the years of the Cold War.[52] Therefore, Bondi felt the need to defend the organizers of the Tbilisi conference and recognize that they had all done their best to solve political problems that arose only after the decision to host the GRG conference in the Soviet Union had been taken. For Bondi, the fact that invitations to Israeli scholars had not been sent was a sufficient reason to withdraw the sponsorship of the ICGRG, and, as President, he would not go as a demonstration of this principle. However, this was not to be taken as a reason to boycott the conference, and even less so was the Czech crisis. The rationale was that if one allowed scientific meetings to be used as an arena for political activities than "indignation over Viet Nam would make out the U.S.A., indignation over atom tests would make out France, over Nigerian arm sales would make out the U.K., and we would have to hold our meetings in the outer space." For these reasons, he publicly stated that he opposed any political boycott of any scientific meeting and publicly regretted Rosen's letter sent a year earlier in which Rosen stated that he would not go to the Tbilisi conference.[53]

---

[50]"[*L*]*a science doit être placée au dessus de la politique*." Fock to Mercier, 11 November 1968, BOND, folder 4/4A, Fock's emphasis.

[51]Bondi to Fock, undated handwritten note, BOND, folder 4/4A.

[52]A quite relevant case here was that the entry visa was denied to delegates of Western European countries that were or had been members of communist parties and wanted to participate in the 16th IUPAC Conference held in New York City in 1951 (Fennell 1994, p. 99).

[53]Bondi to the members of the Committee on GRG, "The events of summer 1968," undated handwritten note, probably 4 September 1968. It is likely that it was actually sent to the ICGRG members. See Bondi to Mrs. Browne, 4 September 1968. See also Bondi to Mercier, undated handwritten note, BOND, folder 4/4A.

Bondi, who had just become the Director General of the much bigger and more important European Space Research Organization (ESRO) in 1967, had formed an accurate, realistic view of the inherent political issues embedded in the functions of international institutions, in the same fashion as the Director General of UNESCO Luther Evans had understood the role of UNESCO in the mid-1950s (see Chap. 3). The only way to solve these issues was to limit the range of inadmissible political matters as far as possible to those officially defined by the most recognized international scientific institutions; namely, ICSU and the international unions with their policies of "Freedom of Scientific Research" and "Policy of Political Non-Discrimination," which had been formally adopted since the IGY.[54] This view was clearly closely related to the opinion that the function of international scientific institutions was only to promote science, and should not involve other more subtle ethical and moral concerns.

The scientists who did not change their plans to attend the conference explicitly or implicitly agreed with the lines between politics and science that Bondi attempted to draw.[55] Strikingly, a number of American scientists, especially those close to Wheeler's circle, did not show any willingness to boycott the meeting. Evidently, Wheeler, who was known as a political conservative and had himself worked on American military programs, saw scientific meetings the same way as Bondi did. He did not allow political matters to get in the way of scientific discussions and advances. Wheeler and the group of scientists more closely connected to him benefited to a great extent from participating in this meeting, particularly in connection with Wheeler's sustained attempt to strengthen the physical character of the GRG domain as well as its links with other branches of physics. This was his main target, and his decision to attend the Tbilisi conference was in keeping with this goal. Bondi's and Wheeler's realistic attitude toward international relations in science could well be considered to be an example of an autoletic "mode of scientific internationalism" to borrow Aant Elzinga's definition, namely, the view that the goal of international institutional activities was to "serve science as an end in its own right" (Elzinga 1996, p. 3).[56]

Mercier had a quite different view—idealistic, if not sentimental—of the role of international scientific community building and of the ICGRG in particular. His hope that scientific endeavors and scientific relations could be a way of achieving peace had been profoundly hampered by recent events, most especially by the

---

[54]See, for example, the case of ICSU (Greenaway 1996) and the discussions within the IUPAC (Fennell 1994, pp. 195–196). However, even in the larger international unions, the principle and its application continued to be matter of debate and redefinition in that period (Greenaway 1996, Chap. 8).

[55]See, for example, Trautman to Bondi, 3 September 1968; and Belinfante to Mercier, 23 September 1968, BOND, folder 4/4A. In his letter, Belinfante criticizes Mercier's actions and then decided to go anyway because "politics should not interfere with science."

[56]Elzinga drew this distinction in the context of institutional settings, but it is also useful to frame the discussion about individual agency.

Czech crisis.[57] The fact that Mercier's devotion to community building was largely based on the complexity of interconnected philosophical, ethical, and scientific reasons was also the reason behind his actions and explains why, for him, it was essential not to separate science and morals. All his activities as a community builder were based on his belief that he was taking moral action aiming toward world peace.[58]

Finally, there was the problem of the demarcation between principles and action. Bondi stressed the principle of separation of science and politics and the related necessity for no scholar to be excluded on political or other unscientific grounds. This position, however, was and remained at the level of principles all through the events. Ultimately, Peres was unable to attend the conference because his visa was not made available, and Bondi was made aware that this was the case.[59] By the same token, Jordan had been invited to the Tbilisi conference, but he never received a visa from the Soviet authorities.[60] To the best of my knowledge, contrary to what happened in the case of the Israeli scholars, this event did not produce any strong protest from other members of the ICGRG. This confirms that some of its members with realistic views, like Bondi, wanted to uphold the established difference between scientific and political issues that was being constructed in the context of the ICGRG more as a matter of principle than in terms of implementing actual policies, something that remained very difficult to achieve in practice.

Rosen and other non-Israeli Jewish scholars—disturbed by Soviet anti-Israeli policies—tended to frame the entire affair as a demonstration of an anti-Semitic attitude in the Soviet Union. Within this understanding of the entire matter, these scientists felt entitled to vociferously protest that one invitation was not enough to prove that there was no discrimination. Robinson, the major organizer of the Texas symposia, was the most outspoken about this. In a letter to Mercier, Robinson made it clear that the "Soviet anti-Jewish campaign" requested scientists in other countries to take a position by either condemning it or by becoming "by their silence [...] its accomplices." He blamed Fock for having put the ICGRG in this situation by "his uncivilized treatment of [their] Israeli colleagues." Since it was now evident that the exclusion of Israeli scholars had become common practice at international conferences held in the Soviet Union after the Six-Day War, Robinson criticized Bondi's willingness to negotiate and condemned all the "tenacious delaying tactics" Fock had used. Fock's tactics led to the absence of Israeli scholars at the conference but, at the same time, prevented any embarrassment. As far as Robinson was

[57]After Bondi had circulated the report on the problems concerning the exclusion of Israeli scientists, Mercier had replied that he would go anyway. Mercier to Bondi, 16 July 1968, BOND, 4/4A.

[58]In Elzinga's definition, this could correspond to the "heteroletic" mode of scientific internationalism (Elzinga 1996, pp. 3–4).

[59]Peres to Bondi, 8 September 1968, BOND, folder 4/4A.

[60]Documentation on the invitation to Jordan are in VFP. For a historical analysis of these events, see Jean-Philippe Martinez, Ph.D. dissertation on Vladimir Fock prepared at the University Paris 7 —Paris Diderot, to be defended in 2017.

concerned, Fock was able to accomplish his task: "the meeting was held without Israelis, but under the auspices of the International Committee on General Relativity and Gravitation." As Rosen had also done, in light of this discrimination, Robinson proposed that the next meeting be held in Israel and suggested that Fock "spare himself and all of us much embarrassment by a prompt and graceful resignation [from the presidency of the ICGRG]," in view of the fact that he had been unable to "abstain from gross discrimination against our Israel colleagues."[61]

The radically opposed positions on what had happened and the boundary between admissible and inadmissible political intervention in scientific international relations were not the only matters open to political discussion. Inherent to the political sphere were also the decisions taken during the reduced Tbilisi ICGRG meeting about the rules regulating the substitution of ICGRG members. Some of those ICGRG members who did not attend the Tbilisi conference reacted very negatively to this motion. They officially requested to postpone the decision to a later meeting of the ICGRG when, hopefully, the majority of members would be present.[62] Bergmann made it clear that a decision of that type did not only concern an inoffensive bureaucratic rule, but had deeper implications on the role and the structure of the ICGRG. For Bergmann, "[a]s long as the members of the International Committee are not elected by relativists at large, the Committee has no legitimate function beyond the one of sponsoring and supervising international conferences. If there is a desire to expand its functions then an international society of relativists ought be [sic] formed with membership open to all which elect its governing council and its officers. If this were to pass, the entire Committee would, or course, dissolve itself and pass in its present function to the new organization. Until then we should avoid all action that would tend to fragment the present International Committee into national delegations."[63]

Bergmann's was the clearest response to the opposing pressures aimed at structuring the ICGRG in different ways. An increasing number of younger scholars were rapidly becoming authoritative experts on GRG in their own right. In keeping with the general climate of generational and political renovation resulting in the protests of 1968, beginning with the London conference, younger scholars—mostly belonging to the American community—were pressing for a transformation of the ICGRG into an open democratic scientific society whose leadership was to be elected by its members (Mercier 1979).[64] Soviet scholars, on the other hand, preferred to maintain the form of the International Committee, but modified so that it better resembled the various international committees of IUPAP. In other words,

[61]Robinson to Mercier, 5 April 1969, ESP, Box 5, folder Ivor Robinson.

[62]Telegram from Alfred Schild to Fock, undated; Bergmann to Fock, draft of an undated letter, PBP.

[63]Bergmann to Fock, draft of an undated letter, PBP; it is unclear whether or not Bergmann actually sent this letter.

[64]In a personal communication, Joshua Goldberg similarly stated that the ICGRG was increasingly seen by the younger scholars as a "self-appointed group" of experts without any right to administrate the GRG international community at the institutional level (see also Held et al. 1978).

they were willing to modify the ICGRG to make it an expression of national delegates on whom the Academy of Sciences of the USSR might exercise a stringent control. Apparently, this request was also related to attempts made by Fock and other Soviet Committee members to have Ivanenko excluded by the ICGRG and substituted by Ludvig D. Faddeev through the decisions made at the Tbilisi conference.[65] Other members of the ICGRG preferred not to take a stance on this. Rosenfeld, for instance, wrote to Bondi that "we ought not to let ourselves be dragged into a purely internal squabble between our Russian colleagues." In his view, the problems had to be settled by Soviet physicists without any external intervention.[66] Ultimately, the decision to substitute Ivanenko with Faddeev was not accepted by the ICGRG at large.

These different views on how the institutionalization of the fervent community of scholars involved in the field of GRG should unfold had been developing for a few years. Until 1967, the ICGRG had successfully managed to promote scientific cooperation in spite of Cold War tensions. The onset of combat—the Six-Day War and the armed invasion of Czechoslovakia—and their impact on the Tbilisi conference had shown that attempts to institutionalize international scientific cooperation in dramatic political situations could not avoid having explicit political implications. Only at that point did the different views on the future of the ICGRG emerge and become part of the open debate, and only then could these different views about the organization of a small institutional body be explicitly understood for what they really were: deep clashes linked to opposing political and cultural systems, which up until then had been lurking below the surface of what was defined by the participants as purely scientific activity.

Bergmann was conscious that the tensions between ICGRG members resulting from the Tbilisi affair did not help in the discussions about the form the ICGRG should assume. Because of the "severe strains" posed on the international collaboration, Bergmann asked the members of the ICGRG "to do everything in [their] power to preserve the international character of [their] scientific endeavors and to avoid all steps that might contribute to the aggravations that will be borne in on [them] by events in the sphere of international politics."[67] To this end, he proposed to "reserve for a calmer time those [items of the minutes of the GR5] that might be considered controversial."[68]

The desire to be constructive and to continue the experience of the ICGRG—notwithstanding the many tensions and disagreements—was predominant in the months following the Tbilisi conference. Mercier had offered twice to resign in view of the strong disapproval of his action from both the outgoing and the

[65]Minutes of the ICGRG during the GR5 conference held in Tbilisi on 12 September 1968, PBP.

[66]Rosenfeld to Bondi, 19 November 1968, BOND, folder 4/4A.

[67]Bergmann to all members of the ICGRG, including Felix Pirani, undated, probably November 1968 ca., PBP.

[68]Ibid.

incoming Presidents.[69] Both of them refused, instead thanking Mercier for his services despite the deep disagreement.[70] Mercier's commitment was considered essential to the successful continuation of the ICGRG enterprise.

Another sign that there was the desire to continue and even strengthen the role of the ICGRG in the promotion of research in GRG was that, in 1969, Germany finally joined the committee. Scientists in the Soviet group decided that the time was ripe for introducing East German theoretical physicist Hans-Jürgen Treder into the ICGRG. A pupil of Papapetrou in East Berlin, Treder acquired a high level of authority within the East German research system coordinated by the German Academy of Sciences at Berlin (DAWB) after Papapetrou moved to Paris in 1961 (see Appendix A.5).[71] In 1963, Treder became Director of the Institute for Pure Mathematics of the DAWB, where Papapetrou had worked before deciding to leave East Germany. Treder's scientific rank, combined with his political attitudes, made him the leading figure in East Germany's GRG research domain, apart from a rival group at the University of Jena, under the leadership of theoretical physicist Ernst Schmutzer, which was established in the early 1960s.

In November 1965, for the fiftieth anniversary of Einstein's formulation of the theory of general relativity in Berlin, Treder had organized a major and high-ranking international symposium entitled "Origin, Development, and Perspectives of Einstein's Gravitation Theory." Sponsored by the DAWB, the event hosted sixty scholars from different Eastern as well as Western countries, including highly reputed experts such as Bondi, Fock, Ivanenko, Lanczos, Mercier, Møller, and Wheeler (Mercier 1966; Treder 1966). The rivalry between Treder and Schmutzer is evident from the fact that Schmutzer organized a parallel event in Jena the same year and neither Schmutzer nor Treder invited the other as a speaker at their conference, a pattern that was to continue in the following years. The Jena conference, however, was more like a workshop and much less prestigious and ceremonial as it was almost exclusively dedicated to active research by scientists working in Germany (especially the GDR, but there were also a few from Hönl's group in West Germany) and Eastern European countries. The only speaker belonging to Treder's group was Georg Dautcourt, who had also been one of the early students of Papapetrou (Anon. 1965).

The larger and internationally renowned Berlin conference was extremely successful launching Treder as the undisputed leader of gravitational research in East Germany. A few months after the conference, he was given the directorship of a new institute devoted to research in relativity and extragalactic physics in Potsdam-Babelsberg (Anon. 1967) and became a prestigious member of the DAWB. In 1969, in the context of the general reform of the DAWB, the growth of

[69]Mercier to Bondi, 27 August 1968; and Mercier to Fock, 3 October 1968, BOND, folder 4/4A.

[70]Bondi to Mercier, 2 September 1968; Bondi to Mercier, 11 October 1968; and Fock to Mercier, 11 November 1968, BOND, folder 4/4A.

[71]For Papapetrou's role in launching research in the GRG field in East Germany, see Hoffmann (2017).

the field was reflected by the foundation of the Zentralinstitut für Astrophysik (ZIAP). ZIAP was part of a larger, newly established administration unit of the Academy called *Forschungsbereich Kosmische Physik*, which, besides ZIAP, included three other institutes mostly devoted to research in geophysics.[72] Treder's scientific prestige in the field of GRG and his strong involvement in science administration in the GDR earned him the position of Director of ZIAP as well as the even more powerful status of Head of the *Forschungsbereich Kosmische Physik*. Both of these reflected the fact that, by the late 1960s, Treder was an eminent member of the East German scientific and political nomenclature.

From the perspective of the Soviet scientists, Treder had certainly gained the authority to become the first East German candidate for ICGRG membership. It was clear, however, that participation in an international organization of this kind—sensitive to the Cold War political balances—required Treder's name to be put forward together with an authoritative representative of the West German community. As Jordan was evidently not to be taken into consideration for the reasons discussed in Sect. 4.2, Soviet scientists decided to propose Otto Heckmann.[73] Eminent astronomer and cosmologist, the long-term director of the Hamburg observatory had also acquired a considerable expertise in matters of international scientific institutions, firstly as Director of the European Southern Observatory (ESO) since 1962 and, later, as President of the IAU from 1967. Moreover, Heckmann was certainly one of the major experts in general relativity of the older generation and held strong links with the group of younger West German theoretical physicists trained by Jordan at the nearby University of Hamburg (see Appendix A.3.1). Despite some opportunistic attitudes during the Nazi regime, Heckmann was considered the ideal choice to counterbalance the entrance of Treder from East Germany.[74] In 1969, Fock made the official proposal that Treder and Heckmann should join the ICGRG, which was accepted by the other members.[75] This decision overcame what had so far been a major exclusion due to the unsettled political status of divided Germany. In parallel to the two Germans, the ICGRG also acquired the first Indian member: physicist and mathematician Prahalad Chunnilal Vaidya (see Appendix A.6).[76] The enlargement of the ICGRG to include Indian and

[72]For the general reform of the German Academy of Sciences at Berlin, which will be also renamed Academy of Sciences of the GDR in 1972, see Laitko (1999).

[73]Georg Dautcourt, personal communication. According to Dautcourt, the idea of proposing Heckmann as Treder's West German counterpart was suggested by him during one of his visits to Ivanenko's group in Moscow. Jordan was not acceptable because of his Nazi past and his former political activities as member of the West German Parliament. However, Dautcourt recalls that opposition to Jordan was not specific of Eastern scholars, but also came from Western scientists.

[74]For Heckmann's career during the Third Reich see Hentschel and Renneberg (1995).

[75]Treder to E.A. Lauter, 30 September 1969, Hans-Jürgen Treder Papers, BBAW, folder 101.

[76]The entrance of Vaidya was probably related to the establishment of the Committee on Gravitation in India, on the occasion of the Research Seminar on Relativity, Gravitation and Cosmology held at Gujarat University, Ahmedabad in February 1969, which was attended by the Soviet ICGRG member Ivanenko. Vaidya was the Vice-Chairman of the newly established Indian Committee (Ivanenko 1969).

German scientists was moving in a direction of greater representativeness, which was being requested by the GRG community at large. These changes had all been promoted by Soviet representatives, probably in an attempt to extend their influence in the ICGRG in a period of crisis and to mitigate the request for more drastic changes coming from scholars of a younger generation, particularly in the United States.

5.2 The Establishment of a New Scientific Periodical: *General Relativity and Gravitation*

Once Fock had rejected his resignation, Mercier re-immersed himself completely in his role of Secretary of the ICGRG with the desire to make its activities even more influential in redefining the GRG field. He embraced and vigorously pursued an idea advanced by the late Infeld: the establishment of the first scientific periodical entirely dedicated to the publication of papers in the field of GRG. In 1962, in a period of rapid growth of the field and with the ICGRG just founded, Leopold Infeld argued that the time was ripe for launching a periodical that would publish exclusively research papers on topics related to GRG. The *Bulletin of GRG* was certainly a very useful tool to challenge the dispersion of the literature and spread important information among scholars working in the field, but, for Infeld, it was not enough: what was needed was to create a truly specialized scientific periodical for publication of new research in the emerging field (Mercier 1979, p. 181).

In general, the need felt by Infeld in 1962 to concentrate the literature in a single publishing venue as a way of challenging the epistemic dispersion of the field and its lack of disciplinary definition was in line with similar attempts to establish comprehensive forms of communication channels regarding new advances. The tension between the growth of the field and its increasing institutionalization, on the one hand, and the still evident dispersion of related research activities, on the other, resulted in two edited books, both published in 1962: *Recent Developments in General Relativity* (1962), a volume in honor of Leopold Infeld, and *Gravitation: An Introduction to Current Research*, edited by Louis Witten (1962), the head of the theoretical research group in gravitation theory at the RIAS in Baltimore (see Appendix A.14.6). The two books had a similar goal: to present a broader view of the advances in the field of GRG without any attempt to build an "orthodox" perspective because many of the topics addressed were still considered controversial, as some of the reviewers emphasized (see, e.g., Bergmann 1962, 1963; Kilmister 1963). The most important and influential of these two volumes was certainly *Gravitation* (Witten 1962). It contained a series of review articles on various research topics—"pedagogical in style"—written by early career scholars.[77] This volume was to be understood as a form of self-contained textbook on recent advances, but it did not have the presumption to present any uniform view of the

[77]Louis Witten, e-mail to the author, 1 December 2016.

GRG field. This modality of communication, in pedagogical form, about new advances suggests that the community at large still felt that the field was epistemically dispersed in different sub-branches and that the common ground had still to be identified.

Notwithstanding the general agreement about the issue of the dispersion of the research field, which resulted in the creation of pedagogical edited volumes, Infeld's 1962 proposal to create a GRG periodical was received with considerable opposition within the ICGRG. Consequently, he decided not to pursue this idea.[78] Although we do not have any documentation that allows us to identify the reasons why Infeld's idea was disputed, it is very likely that those who openly opposed the existence of "relativists" separate from other physicists, like Wheeler, could have seen the establishment of this type of journal as a step in the direction they did not want to move in. Wheeler, together with other physicists with the same attitude as him, might well have seen the idea of a periodical completely and solely dedicated to GRG as the preserve of "relativists." This would have damaged, rather than strengthened, relationships with other sub-disciplines of physics as well as communication with the physics community at large.

Appreciating, however, the necessity to pursue coordinated editorial strategies for consolidating research in GRG, during the 1965 ICGRG meeting, Wheeler proposed publishing a series of monographs on GRG topics through Princeton University Press. It is possible that, in Wheeler's view, this venture might also have served the purpose of defining standards for the entire community, which was another of Wheeler's main areas of focus in his participation in the ICGRG.[79] This proposal, however, clashed with a similar project already begun around 1963 by Cambridge University professor Dennis W. Sciama, a GRG expert of a younger generation who had already become one of the major authorities in the field (see Appendix A.13.1).[80] Apparently, under Sciama's influence, Cambridge University Press had already made plans to publish a monograph series about relativity. As Sciama was not in the ICGRG, this meant that in order to avoid unnecessary competition and replication, it was necessary to coordinate with Sciama, and more in general, with the GRG community at large. In 1965, the ICGRG left the issues concerning the concentration of scientific findings and coordination of publication

[78] Minutes of the Meeting of the ICGRG, 30 June and 7 July 1965, PISGRG, folder 1.1.

[79] Minutes of the Meeting of the ICGRG, 30 June and 7 July 1965, PISGRG, folder 1.1. Back in 1955, Wheeler wrote: "[Bryce DeWitt and Cécile DeWitt-Morette] propose to do something that has long needed doing—help make clear the fundamental facts and principles of general relativity so clearly and inescapably that every competent worker knows what is right and what is wrong. They can do much to clear away the debris of ruined theories from the rocklike solidity of Einstein's gravitation theory so its meaning and consequences will be clear to all." Wheeler to Agnew Bahnson, 25 November 1955 (quoted in Rickles 2011, p. 14). In 1968, Wheeler would reiterate the need to define standards in the field of GRG proposing to "select a standard set of sign conventions, to be used wherever possible by those working in the field." Misner, Thorne, Wheeler, Open letter to Relativity Theorists, 19 August 1968, PBP.

[80] Schild to Drs. Burdine, Hackerman, Hanson, Ransom, Stone, Whaley, 27 May 1963, ESP, Box 3, folder University of Texas.

activities unsolved. The Princeton plans were not pursued, and the Cambridge monographs series began much later and with a more general focus as the *Cambridge Monograph on Mathematical Physics*.[81]

By the time of the ICGRG crisis, there was no coordinated approach to solving the problem of the dispersion of the GRG literature that Infeld had felt so strongly about in 1962 and continued to feel until his death a few months before the Tbilisi conference in 1968. According to Mercier, before Infeld died, "he gave [Mercier] orally a sort of commission, not to drop the idea of a Journal in spite of the reluctance shown by several members of the Committee" (Mercier 1979, p. 180). Approaching the matter with the same sentimental attitude he had shown in his services to the GRG community, Mercier felt the duty to realize what he later symbolically depicted as Infeld's last will for the GRG field. And Mercier did so at a time of major stress for the ICGRG, when the Tbilisi crisis was still very present in the minds of ICGRG members as well as of many scientists in the larger GRG community.

Instead of waiting for the next ICGRG meeting, in May 1969, Mercier sent out a questionnaire to the 26 members of the ICGRG and to more than 200 other individuals on the list of scientists active in the field of GRG published in the issues of the *Bulletin on GRG*. In his letter, Mercier asked for responses to questions concerning "the establishment of a new international JOURNAL on GENERAL RELATIVITY AND GRAVITATION."[82] The explicit motivation behind this proposal was to address the dispersion of information on new knowledge products in the field, which were still scattered in many different journals. His activity as the editor of the *Bulletin on GRG* put Mercier in the position of having a clear idea about the dispersion of the field: The approximately 700 papers that had appeared in the previous 12 months had been published in more than 60 different journals printed in various countries, some of which were not easily available to the majority of scholars involved.[83] Implicitly, however, the proposal had a much greater significance than that explicitly addressed in Mercier's letter. If accepted, it would have had a far-reaching impact on the function of the ICGRG, for the committee would have become the organizational body in charge of the first scientific periodical dedicated to what had by then been identified as the field of

[81]The first book in the series was the important monograph by Hawking and Ellis (1973). The other books were published from the 1980s onward and only a few were on topics related to GRG: *Cambridge Monographs on Mathematical Physics*, https://www.cambridge.org/core/series/cambridge-monographs-on-mathematical-physics/B5B9D3A75391E59CF00429DF1A92AF65. Accessed 10 March 2017.

[82]Mercier to Scientists throughout the world who work in the field of GRG, 27 October 1969, PBP.

[83]Similar concerns had been at the basis of various debates on the formation of specialized sub-disciplines of physics with their own publishing venue. See, for instance, the case of solid-state physics (Hoffmann 2013). For a historical study of the formation of the American solid-state physics community, see Weart (1992).

general relativity and gravitation. Plausibly, Mercier believed that the realization of Infeld's idea would have had a positive impact on the field of GRG and, more implicitly, on the international body that had established itself as the main structure for promoting this kind of research, as it could have been a way to keep the community together in the period of maximum tension due to political matters.

The response to Mercier's proposal might be considered as mostly positive if one ignores the fact that less than half of the ICGRG members and only one-fourth of the other scientists actually replied to the questionnaire. Among those who replied, 67% of the ICGRG members and 84% of the others gave a positive answer to the question as to whether a journal specifically dedicated to the GRG field was desirable. Mercier took these replies as a clear demonstration that he should go forth with the project since he had "a majority greater than what any parliament would require."[84] He made immediate plans for production of the new journal that was to be published under the auspices of the ICGRG. The editor of the new venture, needless to say, was going to be Mercier himself who considered the journal as the natural progression of the *Bulletin on GRG*.

Not all the members of the ICGRG shared Mercier's enthusiasm for the project. Nor did they agree with Mercier that the responses to the questionnaire constituted solid evidence that the vast majority of the GRG community was in favor of the enterprise. Bergmann was the most outspoken critic of the entire endeavor. After having discussed the matter with Wheeler and other members of the North-East American part of the GRG community, Bergmann came to the conclusion that many of his colleagues were afraid that the new journal would only "do harm to our field."[85] Although Bergmann was not explicit about the reasons for the opposition, it seems reasonable to suppose that these American physicists feared that the establishment of a journal dedicated to the GRG field would increase the isolation of GRG with respect to other branches of physics.

Since his letter of protest failed to dissuade Mercier, Bergmann asked Bondi to intervene in view of Bondi's close relationship to the "European scene." Bergmann argued that Mercier could start the new journal if he wished, but he should not act in the name of the ICGRG "basing his authorization on a mail vote arranged by himself, in which less than half of the membership voted."[86] Bergmann was very worried that Mercier's decision to publish the journal under the sponsorship of the ICGRG would become a "serious threat to the continuity of the Committee," in a period when the ICGRG was already under acute stress.[87] Bergmann's concerns demonstrate that there was much less agreement on the establishment of the new journal than Mercier was willing to recognize. Furthermore, Bergmann's words confirmed that there were identifiable national, or even local, communities with

[84]Mercier to ICGRG members, 22 September 1969, PBP.

[85]Bergmann to Bondi, 8 October 1969, PBP.

[86]Ibid.

[87]Ibid.

different needs, and consequently different strategies, regarding how the field of GRG should be promoted.

Despite the opposition, Mercier—as in case of the Czech crisis—pursued his own project and his own vision, modifying the ICGRG and its activities for many years to come. Acting under the assumption that the vast majority was supporting his decision, and believing that he was in any case doing the right thing regardless of all the possible objections, Mercier made plans to rapidly publish the first issue of a periodical that would circulate original papers, research reviews, and book reviews in the field of GRG, as well as the kind of information that had hitherto appeared in the *Bulletin on GRG*.[88] To ensure that the journal would start off on a strong footing and become the central publication venue of the entire field, Mercier also asked for direct contributions from the ICGRG members by inviting them to submit first-rate research articles. According to Mercier, this would not only show that the journal contained important new research products, but it would also demonstrate that the ICGRG members were supporting the new project.[89] The first issue of *General Relativity and Gravitation*: *A Journal of Studies in General Relativity and Related Topics* was published in March 1970. Its 101 pages did not contain a single contribution from any ICGRG members. Nonetheless, the journal began to immediately publish significant papers that would be soon perceived as groundbreaking contributions. It rapidly became the most important publishing venue for research in the field of GRG and, a posteriori, the periodical was considered to be Mercier's greatest achievement in GRG.[90]

5.3 Toward the International Society on General Relativity and Gravitation

After Møller had confirmed that the next conference could in fact be held in Copenhagen, preparations began for organizing what would become the largest conference on GRG so far: the GR6 conference held in Copenhagen from 5 to 10 July, 1971. The field had grown enormously from the first tentative attempts to build an international community in the mid-1950s. Those who were students in the centers active in the mid-1950s had meanwhile found permanent positions and inaugurated new research centers, where a third generation of scholars who were already contributing significantly to the field were now studying.

The transformation was not only quantitative and not only dependent on the coming of age of the second generation of experts in the field. The field itself had

[88]Mercier to Subscribers to the Bulletin on GRG, 2 February 1970, PBP.

[89]Mercier to ICGRG members, 31 October 1969, PBP.

[90]As some of his colleagues later stated in the pages of *General Relativity and Gravitation* "[f]rom the point of view of the world of relativity, perhaps [Mercier's] most important contribution was the founding of this journal" (Held et al. 1978, p. 760).

completely changed during the 1960s and was still in a state of permanent flux. The discoveries in astrophysics and the growth of new technologies had made relativistic gravity an observational science in every respect. In the 1950s, this status had been very much uncertain, with a few pioneers proudly making the first attempts to design possible tests of general relativity and alternative gravitational theories. In contrast, around 1970, many groups were actively engaged in gravitational tests and obtaining important results confirming Einstein's theory (Will 1986; Wilson and Kaiser 2014).

The most clamorous of these experimental results was in gravitational radiation research. After ten years of work in isolation, Joseph Weber of the University of Maryland had announced in 1969 that he had eventually found convincing evidence that he had detected gravitational waves with his aluminum bar detectors, known as Weber bars. Gravitational radiation had been one of the most studied theoretical topics in general relativity since the Chapel Hill conference in 1957. From the observational side, however, only a few physicists had proposed attempts to observe it. Weber, a former postdoc associate of Wheeler's, was the only one to persistently pursue what was commonly perceived as the virtually impossible task of observing this extremely tiny effect. Weber's announcement in 1969 sparked an enormous amount of excitement. The potentially Nobel-worthy discovery in the field of GRG resulted in a fervent theoretical activity with the aim of understanding the plausibility of Weber's finding, which was seen with skepticism by many scholars. At the same time, the announcement sparked a rapid growth of experimental activity in the domain of gravitational wave research. By the time of the GR6 conference, six experimental research groups were actively pursuing the program of repeating Weber's experiment in the United States, Western Europe, and the USSR (Franklin 1994; Collins 2004; Franklin and Collins 2016; Trimble 2017).

Relativistic astrophysics, observational cosmology, experimental gravity physics, and, finally, experimental gravitational wave research had all made GRG a well-established part of physics. For some of the practitioners, this was the main breathtaking transformation that deserved the title of "Renaissance of General Relativity" (Will 1989).

The return to the mainstream of physics, however, was not only in terms of connections with observational and experimental activities. It also involved radical transformations in theoretical practices and ontologies. From the conceptual perspective, the implications of the theory had given rise to non-Newtonian entities, like black holes and gravitational waves, which were now being accepted by the majority of physicists as (testable) elements of the physical world. In terms of practices, the increasing relevance of observational and experimental activities changed the priorities for theoretical research as well. Within the physics discipline, it became much more valuable to pursue theoretical research in general relativity when the links to the empirical world were clear, which then led theorists to collaborate directly with experimental groups and perform the calculations that were most conducive to advances in the experimental range. The recent transformation of the status achieved by general relativity around 1970 was, in a certain sense, the fulfillment of the initial dream of Wheeler's who had envisaged a full integration of gravitation theory into the physics domain, as opposed to the formation of a separate field of "relativists" (Blum 2016). It comes as no surprise that some of his early associates, such as Charles Misner, Kip S. Thorne, and Weber, were playing a particularly prominent role in this process.[91]

Since 1959, when the ICGRG was established, the GRG community at large had undergone a radical change of both social and epistemic nature. Yet the composition of the committee had remained almost the same. The initial members had tried to deal with the ongoing transformations by enlarging the membership, but the core of the ICGRG had not been modified and the new entries could not be considered as a fair representation of the radical changes that had occurred meanwhile. Many of the major figures who emerged in the 1960s were excluded by the decisions of the ICGRG and, by contrast, some of the ICGRG members were no longer very active in the field. This was at the basis of a diffused discontent among the younger generation of GRG experts who did not feel that the committee fully represented them and their research interests.

No doubt, most of those involved in this transformation looked favorably at the formation of the field of relativistic astrophysics. These actors saw the highly attended biennial Texas symposia as the place where new exciting knowledge was circulated, partly thanks to the opportunity for younger scholars to openly discuss their research with more established ones without a strong hierarchical structure. It is plausible that the Anglo-American scientists who appreciated the Texas symposia found the GRG international conferences less exciting and somewhat backward in comparison. Although in his early days of building an international community of relativists Mercier had in principle aimed to resurrect general relativity as a physical theory, he held completely different views about how this should have occurred. In the pages of the *Bulletin on GRG*, Mercier did not refrain from criticizing the format and the contents of the Texas symposia. Mercier began by deploring "the lack of good style, rhetoric and good didactic of presentation from many a speaker who seem to fall into the (American?) habit of presenting their subject-matters very carelessly" (Mercier 1965, p. 12). Mercier's comment on the Third 1967 Texas Symposium with about 600 participants was even more critical. He wrote that the large number of short communications, often without clear conclusions, as well as the lack of a clear structure according to research topics made it difficult to form any understanding of what were the main findings in the field. For Mercier, this shortcoming meant that the event was more like a gathering of a "lose [sic] Association," than a real conference. He even went so far as to dispute the use of the adjective "relativistic," declaring that it was a purely astrophysics event in which the status of the field was so uncertain that no synthetic view could be achieved.[92]

---

[91]As confirmation of this trend coming especially from Wheeler, one might notice that the move toward using supercomputers for the solution of Einstein's equations was again related to what philosopher and historian of science Dennis Lehmkuhl (2017) calls Wheeler's family.

[92]"[T]he adjective Relativistic, used in the invitation, surely did not apply" (Mercier 1967, p. 10).

The contrast between the GRG community that had emerged in the last decade and the one represented by the ICGRG was symbolized by the completely different formats of the Texas symposia and the GRG international conferences. In terms of topics, the evolution of the field was taken in due consideration in the organization of the GR6 Conference in Copenhagen. For the first time, the schedule of invited lectures was prepared with the aim of maintaining the balance between observational and theoretical matters, often in combination.[93]

In an operative meeting of the ICGRG in Gwatt, Switzerland, one year before the GR6, Wheeler once again pushed for the realization of his view of the conference, which he believed should be based on the principles of "breadth," "excellence," "physical relevance" and "newness," wherein with *physical relevance* Wheeler meant "tied to the real world by some very fairly clear line of reasoning, (as distinct for example from one more solution of field equations.)."[94] Even though some members of the ICGRG could, at least in principle, agree with Wheeler's list of the priorities that should govern the organization of the next conference, others moved in different directions in terms of the topics to be addressed.[95] Wheeler's views of the field, its priorities, and its direction were, most likely, shared by the majority of physicists now participating in the venture and were gaining momentum. Nevertheless, given the specific composition of the ICGRG and its perceived role as being representative of the various local communities, the tension between different kinds of approaches within the ICGRG was still high.

The decision concerning the topics was directly related to the choice of individuals to be invited at the conference. At Gwatt, the ICGRG members decided to limit participation to 200 attendees. Although there might also have been practical reasons for putting an upper limit on the number of participants, the main rationale was again that the organizers and the majority of the ICGRG members considered conferences with a limited number of participants most productive in terms of discussion and scientific progress.[96] With this decision, they were following the policy promoted by Møller and Rosenfeld—two of the four organizers of the conference—back in 1961 when they proposed limiting attendance at these conferences to around 100 participants (see Sect. 4.4). For Rosenfeld, in fact, the limitation was essential in order to "provide favourable possibilities of fruitful

[93]Letter of Invitation, GR6P, Box 1. See also The Organization Committee (Møller, Rosenfeld, S. Rozental, B. Strömgren) to the members of the ICGRG, 24 June 1970; and Fock to Møller, 12 August 1970, GR6P, Box 1.

[94]Wheeler to Strömgren, Møller, Rosenfeld and Rozental, 18 May 1970, handwritten at Gwatt Switzerland and given to Rozental, GR6P, Box M-Z.

[95]While Wheeler stressed that new results might come from relativistic astrophysics, Bergmann, for instance, criticized the preliminary schedule as being "slanted toward the astrophysical and cosmological aspects," while he suggested giving more space to other kinds of theoretical advances. Bergmann to Møller, 20 July 1970, GR6P, Box 1.

[96]Møller to A. Fischer, 23 February 1971, GR6P, Box A-L.

discussions,"[97] in opposition to "a variety of examples (the worst being the High Energy fairs, vanity fairs in fact)."[98]

The problem was that while 100 participants might have been a fair number for a small, emerging field such as GRG in 1961, by 1971, 200 people had become far too small a number considering all those active in the field.[99] The organizers had to reject a significant number of requests from scientists they had not invited.[100] Many complaints were made by scholars who hoped for a greater level of openness at these international conferences. These complaints did not come only from those who were excluded. Invited authoritative scientists openly criticized the ICGRG's decision to limit participation at the conference to 200. One of the most vocal critics was Felix Pirani. He lamented that this policy would establish a "hierarchical tradition of invitations" that precluded the participations of younger scholars, who were, in his opinion, those who would have benefited most from events of this kind.[101] For the same reason, Arthur Fischer of the University of California at Berkeley openly accused the conference of an "elitism […] which [was] alien to an international community of scientists."[102] Fischer was also reporting the dissatisfaction of a number of young scientists he had met at the recent Texas symposium, held in December 1970, where there was "a great deal of bitterness by the prospect of being not allowed to attend the conference."[103]

The large number of people excluded from the conference and the open opposition to the invitation-only policy led a section of the community to question once again the authority of those in charge of deciding who was worth being invited to these exclusive socio-epistemic events. The ICGRG was responsible for all these decisions, but no one had elected its members who had such strong power to include or exclude people, and to decide which topics and research findings should be discussed at institutionalized events. These events could only increase the ongoing conflict between the younger generation working in the field and the members of this committee established twelve years previously when radical changes in the discipline had yet to occur. Together with the political controversies surrounding the Tbilisi conference, the debate on the invitation-only policy for GR6 gave renewed impetus to the long-standing request for a radical transformation of the ICGRG that was coming from a large section of the GRG community, mostly related to the American sphere and relativistic astrophysics.

---

[97] Rosenfeld to Gerald Tauber, 19 April 1971, GR6P, Box M-Z.

[98] Rosenfeld to Felix Pirani, 19 November 1970, GR6P, Box M-Z.

[99] "[T]he number of active relativists has now become so large that we found it impossible to satisfy all of them," Rosenfeld to Peter Rastall, 19 February 1971, GR6P, Box M-Z.

[100] In May 1971, Stefan Rozental stated that they had to send rejections to more than 50 people. Rozental to Jakob Bekenstein, 27 May 1971, GR6P, Box A-L. Probably, the number of those who wished to attend, but were unable to, was even greater because many younger scholars did not ask directly. See, for example, Pirani to Rosenfeld, 6 January 1971, GR6P, Box M-Z.

[101] Pirani to Rosenfeld, 15 October 1970; and Pirani to Rosenfeld, 12 May 1971, GR6P, Box M-Z.

[102] Fisher to the Organizing Committee of GR6, 12 January 1971, GR6P, Box A-L.

[103] Ibid.

The request to move in the direction of a greater democratic structure was embraced by Bergmann and DeWitt. They became the spokespersons for large sections of the GRG community, dissatisfied by the representative level of the ICGRG. About two months before the GR6 conference, DeWitt sent a letter to the ICGRG members requesting a discussion and vote on a 12-point motion for radical changes to the ICGRG during the next meeting in Copenhagen. DeWitt's proposal was to enlarge membership to 60 scientists and to establish a rapid process of rotation beginning with the immediate dismissal of the present ICGRG members. Among these 60 new members, ten scientists should be elected to form the Executive Council, in charge of organizational matters. As for geographical or national representation, which had been one of the most controversial topics in the previous discussions concerning the composition of the ICGRG, DeWitt proposed that the new committee could decide to establish regional quotas, but the power to elect members was not to be delegated to any region.

In his letter, DeWitt succinctly outlined why he deemed it necessary to approve his plan and to do so as soon as possible:

> 1. The time for action can no longer be delayed. 2. It is necessary to have a complete new scheme in operation before the end of the Conference; otherwise action will only be postponed until the next meeting (an all too familiar pattern). 3. It is impossible to enfranchise all of the physicists around the world who are interested in general relativity; the boundary of such a group can never be defined with precision and, in any case, to ask you to poll them all would be to ask you to accept an impossible burden. 4. By increasing the Committee membership to 60, however, many younger physicists would be brought into the group, and a much more nearly representative world body would be achieved. 5. The device of annual rotation and limited tenure ensures, moreover, that nearly everyone of statute will ultimately have a crack at Committee work. 6. An elected Executive Council will be a much less unwieldy body than the present Committee and much more capable of getting things done. 7. Finally, any of us who fears loss of prestige or privilege by being dumped from the Committee ought to accept the fact that if his prestige and privileges depend *that* much on Committee membership, he ought to be dumped![104]

DeWitt's proposal, supported by Wheeler, was discussed at the first of the three ICGRG meetings held during the Copenhagen conference, on 5 July 1971. As expected, scientists from the Eastern Bloc countries strongly opposed such a radical modification of the ICGRG. For Ivanenko, the proposed change "would be suicide for the Committee."[105] Treder, who was unable to attend the conference for health reasons, sent a letter to Møller declaring that, for him, the proposal was "unintelligible."[106] In Treder's view, the changing epistemic status of the field over the past decade through a merging of general relativity with other fields of theoretical physics should not lead to a modification of the ICGRG. Since it had become well

---

[104]DeWitt to Mercier, 28 April 1971, GR6P, Box A-L, DeWitt's emphasis.

[105]"Minutes of the Meetings on the Committee on GRG, held on occasion of the GR6-Conference in Copenhagen," p. 4, DAUT.

[106]Treder was ill with chronic pancreatitis. Dautcourt, "Bericht über die 6. Internationale Gravitationskonferenz in Kopenhagen," DAUT; and Treder to Møller, 17 June 1971, GR6P, Box M-Z.

established that general relativity was part of "normal" physical and astrophysical research, there was no need to create a special organization of "relativists." This kind of organization might have made sense if the field had an esoteric character, which, Treder stressed, was fortunately no longer the case. Therefore, he proposed maintaining the structure of an "international scientific club" which found its justification only through the level of international cooperation realized so far and by no means claimed to represent the entire field of "Relativity Theory."[107]

Lichnerowicz partly supported the resistance of the Eastern Bloc colleagues proposing a less radical modification, but for completely different reasons than Treder. He believed that one should recognize that, whatever the criticisms from younger scholars and the complaints of underrepresented national communities, the ICGRG had done important work in having permitted scientists to reach "a certain unity of relativity research [which] must be maintained in future."[108] For him, relativity research had become a unified field, also thanks to the work of the committee, and as such it deserved an organization of its own.

As is evident from the different perspectives expressed during the meetings, the apparently formal regulation of the number and rotation of members had a much stronger bearing for the proponents and their opponents, which also concerned the epistemic status of general relativity as a field of research and its links to other fields. The proponents wanted to establish a completely new institutional body that would represent the new fields of research, which DeWitt and Wheeler believed should have considerably stronger links with other areas of physics, possibly in a way similar to what was happening in the emerging and, according to Mercier, confused field of relativistic astrophysics. Some of the opponents argued that it was better to maintain the character of a club of selected scientists who were able to promote international relations, without any need to enlarge the scope of the ICGRG to include several other branches of research. What was at stake was the status of the field and of its practitioners in relation to other branches of physics and

---

[107]Treder to Møller, 17 June 1971, GR6P, Box M-Z. "Mir scheint, daß die Relativitäts- und Gravitationstheorie in dem letzten Jahrzehnt sich in so breiter Front mit der "normalen" theoretischen Physik und Astrophysik verschmolzen haben und daß jetzt auch für eine erfreulich große Zahl von "normalen" physikalischen und astrophysikalischen geworden sind, daß an sich für irgendeine neue Spezialorganisation der "Relativisten" keinerlei Notwendigkeit mehr besteht (und eine solche spezielle Organisation auch keinen definierten Mitgliederkreis mehr haben könnte). - Eine solche relativistische Organisation war so lange berechtigt und notwendig wie die allgemeine Relativitätstheorie einen gewissen esoterischen Charakter hatte.

Ich halte daher alle Vorschläge, irgendeine neue eigene Organisation für Relativitätstheorie ins Leben zu rufen, für nicht mehr recht zeitgemäß, - Das an sich so angenehme könnte m.E. weiterhin als eine Art "Internationaler wissenschaftlicher Klub" bestehen bleiben, der sich nach Vorbild der internationalen und nationalen Akademien selbst ergänzt und natürlich keineswegs den Anspruch erhebt, die Relativitätstheorie zu repräsentieren (dies könnte m.E. heute keine noch so perfekte Organisation mehr)."

[108]"Minutes of the Meetings on the Committee on GRG, held on occasion of the GR6-Conference in Copenhagen," p. 5, DAUT.

astrophysics as well as, for some of them, of mathematics and astronomy. In a certain sense, the ICGRG had contributed to establishing a field of research whose successes in physics and astrophysics were jeopardizing its very status as a recognizable field. Of course, those who had considered themselves "relativists" since the 1950s, such as Mercier and Bergmann, preferred to maintain an institution that preserved the status of the field as such. With his proposal, DeWitt, and with him Wheeler, were simply trying to renew the institutional apparatus to make it more dynamic and more in line with what was happening in current physics and astrophysics research, without being too concerned about the status of the "relativists," an expression Wheeler had always contested.

It is certainly no coincidence that opposition to DeWitt's proposal came primarily from Eastern Bloc scholars, who gained prestige and authority in their home country by participating in this international venture, especially since GRG had meanwhile become an important scientific field. Ivanenko, in particular, had acquired authority within the power dynamics of Soviet scientists through his participation in the ICGRG as a Soviet delegate. Also thanks to this official recognition, he was able to initiate the Gravitation Committee and became the organizer of the Gravitation Section of the Ministry of Higher Education in the Soviet Union.[109] For this reason, Ivanenko had already fought against the decision of his Soviet colleagues to replace him by Faddeev during the Tbilisi conference and was not willing to accept either the loss of prestige related to the transformation of the committee or his possible dismissal (see Sect. 5.1). In the debate about the structure of the ICGRG, epistemic, political, and personal aspects were intermingled and no effort was made to separate them.

To solve the seemingly irreconcilable positions between those who wanted a larger representation of different research groups and those who wished to maintain a more traditional structure by slightly modifying the committee with a few younger researchers, Bergmann proposed transforming the committee into a full-fledged relativity society in analogy to the American Physical Society. Possibly inspired by the recent establishment of the European Physical Society in 1968, Bergmann had already suggested transforming the ICGRG into a "democratically structured society" in the previous meeting.[110] Now he elaborated on his proposal by arguing that it served two related goals. On the one hand, it allowed the existence of a small committee as the governing board of the society, which would follow the tradition established with the ICGRG. Implicitly, this meant that the level of prestige attached to the ICGRG membership would not have changed. It also meant that

[109]"Dmitri Ivanenko—Scientific Biography" http://istina.msu.ru/media/publications/articles/91d/63b/5382068/Biography-Ivanenko.pdf. Accessed 17 February 2017; see also Ivanenko to Møller 22, July 1971, GR6P, Box A-L.

[110]Bergmann to Møller, 10 June 1971, GR6P, Box A-L. In fact, Bergmann had explored the possibility of making the ICGRG a division of the European Physical Society but desisted because American physicists complained that the European Physical Society was not truly international, but rather regional.

Fig. 5.3 Photo of the reception at the GR6 conference taken by the firm Atelier Bache at the Town Hall (Rådhuset), Copenhagen, 6 July 1971, NORDITA Collection, Niels Bohr Archive. I am very grateful to Helle Kiilerich for having located this picture

those who wished to consider themselves "relativists" would have been certainly invited to do so. On the other hand, the society would make it possible to "intensify the contact among scientists working in the field of pure maths, standard physics or standard astronomy related to general relativity," thus allowing the democratic participation of all those interested in the field.[111] Finally, as a democratic society, it would certainly meet the needs of the younger scholars who had been criticizing the elitism of the ICGRG.

After resolving that it was necessary to modify the ICGRG one way or another, the majority of participants accepted Bergmann's proposal, which was likely considered less traumatic than a complete replacement of the ICGRG with a larger committee. As for the procedure concerning the renewal of ICGRG membership, it was voted that eight members—that is, one-third of the current 24 members—would be replaced every three years. It was deemed necessary to immediately start applying this procedure in order to meet the requests coming from the majority of conference participants, and from those who were unable to attend the event. The choice of the first eight members to be substituted was of course controversial. The

[111]"Minutes of the Meetings on the Committee on GRG, held on occasion of the GR6-Conference in Copenhagen," p. 5, DAUT.

final decision was to leave this delicate matter to fate. A ballot was organized in which lots would be drawn by Mrs. Kirsten Møller, the wife of the new ICGRG President and Chairman of the GR6 conference (Fig. 5.3).

During the next committee meeting, the ICGRG members had to discuss two politically delicate points: the exact procedure for replacement of the eight members and the location of the next conference. They began with the latter. Rosen had again proposed Israel as the hosting country and promised that visas would be made available for every participant, including those coming from countries with which diplomatic relations did not exist.[112]

Vocal objections came from Soviet scientists. Fock maintained that the decision to host a conference in Israel would be a *political* action "since Israel is in a state of war with his neighbours."[113] For Fock, under the principle of separation between science and politics, the ICGRG should abstain from such an action. No one responded to this objection, probably because it had already been established during the discussions concerning the Tbilisi conference that it was not possible to exclude countries because of their local or foreign policies. Ivanenko used a different strategy. He proposed accepting the invitation under the condition that political affairs were settled within the year in view of the difficulties related to the lack of diplomatic relations between Israel and most countries in the Soviet sphere of influence. Implicitly accepting Ivanenko's proposal, Bondi proposed voting on a motion in which the final decision was to be postponed until the next year. While Rosen was in favor, others voiced their opposition, claiming that Israel should have been accepted immediately without any conditions. In the discussion, Rosenfeld pointed out that the ICGRG should be aware that "scientists in Russia cannot behave as free men" in terms of participation in international gatherings as shown by the fact that Zel'dovich and Lifshitz had not been allowed to attend the Copenhagen conference.[114]

Bondi's motion was rejected by only two votes with ten members against, eight members in favor, and two abstentions. Since Ginzburg, Treder, and Trautman were not present at this meeting, one might suspect that had all the ICGRG members been there, the outcome of this vote would have been quite different. The refusal of Bondi's motion led DeWitt to propose accepting Rosen's offer unconditionally, which was approved with thirteen votes in favor, four abstentions, and three

[112]Rosen had officially proposed Israel before the conference. Rosen to Mercier, 24 March 1971, GR6P, Box M-Z. Apparently, this information was already known to Soviet scholars who had received a directive from party authorities to oppose this decision (Khalatnikov 2012, p. 134).

[113]Minutes of the Meetings on the Committee on GRG, held on occasion of the GR6-Conference in Copenhagen," p. 8, DAUT.

[114]"Minutes of the Meetings on the Committee on GRG, held on occasion of the GR6-Conference in Copenhagen," p. 8, DAUT. It seems that also Khalatnikov, whom Rosenfeld did not mention, was not allowed to attend. Khalatnikov was scheduled as a main speaker, but he did not come despite the fact that the organizing committee tried to put pressure on the Soviet Academy of Sciences in order to allow him to participate. Møller to the President of the Academy of Sciences of the USSR, Mstislav V. Keldysh, undated typewritten copy of a telegram, GR6P, Box A-L.

against. (These were certainly the three Soviet ICGRG members, Fock, Ivanenko, and Petrov). Through this vote, Israel was formally selected as the host country for the next conference, to be held in 1974.

Once the problem of the next location had apparently been solved, albeit controversially, the ICGRG went on to discuss the procedure for replacing retiring members. After Fock's protest against the decision to immediately substitute eight members, the ICGRG continued as it had been established. In the meantime, Møller's wife had drawn the lot with the names of the first eight scientists who would retire at the end of the Copenhagen conference. The outcome was unfortunate for the German representatives who had become members less than two years previously, Heckmann and Treder, both being selected, along with two Soviets (Ginzburg and Petrov). Meanwhile, no names of Americans were drawn. In order to maintain the principle of geographical representativeness, which had informally guided choices concerning the composition of the ICGRG since its inception, it was necessary to establish a formal regulation concerning the rotation of membership belonging to specific countries.

Mercier took on the task of developing a method by dividing the GRG community in geographical areas. In his proposal, Mercier was attempting to establish a general geographical classification, but with the clear goal of finding a specific practical solution for substituting the members randomly selected by Kirsten Møller. Mercier divided the community into four geographical regions: a) the U.K., Canada, etc. (for the retiring members, Bondi and Dirac); b) France, Italy, Spain, etc. (for Tonnelat and Cattaneo); USSR (for Ginzburg and Petrov); and German-speaking and Central Europe (for Heckmann and Treder). The U.S. was, of course, not included in Mercier's regions because no American name had been selected. The problem, which the ICGRG members did not discuss, was that the pressure to modify the ICGRG came primarily from physicists working in the U.S., while Soviet scholars would have preferred to maintain the ICGRG structure and regulate the rotation between members through choices made locally by the Soviet Academy of Sciences or other centralized institutions of gravitational research.

In addition to proposing geographical regions, Mercier went so far as to include a list of possible candidates, thus projecting his own personal preferences on the future of the ICGRG. How these new members should be elected remained undecided, however. This decision depended on the structure of the society, which was still to be formed. Two non-members of the ICGRG, Peter Havas and Ted Newman, had in the meantime circulated a document requesting that the decisions of the ICGRG be made public. To meet with the needs of the GRG community at large and to explain the decision to establish a society, Bondi proposed to arrange an ad hoc General Assembly to which all participants of the GR6 conference would be invited. This ad hoc assembly was to make the final decisions.[115]

[115]"Minutes of the Meetings on the Committee on GRG, held on occasion of the GR6-Conference in Copenhagen," p. 10, DAUT.

Before the ICGRG members could discuss the matter with the other participants of the Copenhagen conference, they met one last time a few hours before the ad hoc assembly on 8 July 1971. An 11-point statute of the proposed International Society on General Relativity and Gravitation (ISGRG), as it was soon called, was prepared overnight by Bondi and Mercier and circulated among ICGRG members.[116] At the meeting, Bondi explained that this was only a draft constitution, to be presented at the General Assembly in order to ascertain whether the majority of the scientists attending were favorable to the proposed transformation of the international institution devoted to GRG. Yet it contained many rules that should be implemented soon, although the society would not be formally established until the following international conference. The most important point was the regulation about the election of the eight new members of the ICGRG, which should be pursued at the General Assembly.

Once again, Ivanenko attempted to avoid a situation where the participants of the Copenhagen conference at large chose the new members by requesting the final decision to remain in the committee's hands, while the members of the assembly could only provide advice. Ivanenko's proposal was not accepted. Instead, it was decided to enlarge the number of nominees initially proposed by Mercier alone and to ask the members of the assembly to vote between the names chosen by the ICGRG. The first step of the not yet established society would be to approve the spirit of the constitution, confirm the choices of the ICGRG, and then elect the new members of the ICGRG from among those selected by the ICGRG according to the principle of geographical representation.

A most controversial point was that every participant could vote for each region, thus depriving Soviet central organizations of the chance to choose the representatives they preferred. Another major problem was the process of writing the final constitution. While some people would have preferred the ICGRG to carry out this task, others—more attentive to the pressure coming from non-ICGRG members—proposed that an ad hoc committee be formed with the sole function of writing the constitution. Some members of this ad hoc committee should be chosen by the ICGRG, while the members of the assembly should select the others.

This was the situation when the ad hoc assembly convened at 5 p.m. the same day. The chairman of the assembly was Dennis Sciama, who was the internationally recognized leader of one of the most important groups in the field of GRG, but not a member of the ICGRG. After Bondi presented the spirit of the constitution and major points of the draft constitution prepared by himself and Mercier, Sciama explained the regulations concerning the election to be held during the Copenhagen conference.

The constructive discussion on the new society and its constitution following these two presentations did not last long. While the ad hoc assembly of participants was discussing whether and how to transform the ICGRG into a society, Robinson

---

[116]Mercier to members of the sub-committee on the foundation of the society, 18 May 1972; and Mercier to A.A. Sokolov and N.V. Mitskiévic, 1 February 1973, PISGRG, folder 1.2.

expressed inflamed criticisms about the failure to invite Israeli scholars to the Tbilisi conference.[117] He concluded his strong discourse by asking the General Assembly to "express regret at connecting political actions with scientific affairs by the Soviet Union."[118] Ivanenko attempted to avert this strained situation by asking the Chairman of the General Assembly to stop Robinson's speech and to address more scientific topics instead. Apparently, this did not happen. Feeling under attack in this international setting and under political pressure in their home country, the Soviet scholars abandoned the ad hoc assembly en masse. The situation created a problem for the participants coming from the other Eastern Bloc countries, who did not know what to do next. It seems that most East German and Polish participants followed the lead of their Soviet colleagues and left the room, but it was not a coordinated action.[119] In that precise moment, twelve years of efforts by ICGRG members to establish international cooperation were almost nullified. It was thanks to some reconciliatory gestures made by American theoretical physicist Kip Thorne and Soviet experimental physicist Vladimir Braginsky in particular that it did not come to that. Braginsky decided to rejoin the ad hoc assembly. While discussion had continued following Robinson's direct attack, Braginsky briefly explained about the organization of the Tbilisi conference and urged "the Chairman to bring the Assembly to speak about scientific things."[120] With the aid of some American colleagues, such as Thorne and Wheeler, he attempted to allay the tension.[121] However, some people in the room still wanted to protest about what they considered to be open anti-Semitic policies. Banesh Hoffmann, in particular, stressed that while it was certainly better to avoid bringing politics into scientific affairs, "politics has already entered."[122] Mercier then intervened and tried to interrupt this discussion with a motion in which he asked to proceed with the other points on the agenda. This was accepted with 43 votes in favor and 36 against—a demonstration of how high emotions about the Tbilisi affair were still running among some of those present during the assembly.

Once the discussion could continue, the members of the assembly explicitly requested to elect at least one new American delegate and, therefore, add the U.S. as

[117]Georg Dautcourt to Møller, 29 October 1971, PISGRG, folder 1.3.

[118]"Minutes of the 1st Meetings of a General Assembly towards the foundation of Society for General Relativity and Gravitation GRG," p. 3, DAUT.

[119]Ivanenko to Møller, 22 July 1971, PGR6, Box A-L; Dautcourt, "Bericht über die 6. Internationale Gravitationskonferenz in Kopenhagen," DAUT. According to Dautcourt, for instance, he remained in the room. Dautcourt, personal communication. Other recollections seem to imply that every scientist working in Soviet Bloc countries had to leave the room. Genot F. Neugebauer, personal communication. If Dautcourt stayed, it was certainly perceived as a strong political act.

[120]"Minutes of the 1st Meetings of a General Assembly towards the foundation of Society for General Relativity and Gravitation GRG," p. 4, DAUT.

[121]Dautcourt, personal communication.

[122]"Minutes of the 1st Meetings of a General Assembly towards the foundation of Society for General Relativity and Gravitation GRG," p. 4, DAUT; and Kip Thorne, personal communication.

the fifth region, against the decision by the ICGRG to select retiring members by a random lot. Following initial resistance, the pressure of the assembly was so great that DeWitt and Wheeler spontaneously retired in order to allow the election of two new American scientists in the ICGRG.

Another regional matter was also troubling. The fourth region designated by Mercier included German-speaking countries without any consideration of the political division. Most likely under consultation with the Soviet scientists outside the room, Wheeler proposed dividing the fourth region into West Germany (plus neutral Austria) and East Germany (plus the other Eastern European countries) so that the political balance would be respected. Braginsky strongly supported the proposal, while Mercier preferred his own option that would have allowed them "to keep a certain homogeneity in Europe."[123] The assembly agreed with the political separation and accepted Wheeler's proposal. In doing so, the participants were also accepting the requests put forward by Soviet Bloc scientists to implicitly follow the logic of national representativeness. Retiring members were replaced by new members either of the same nationality or belonging to the same wider geopolitical area.[124]

While the general rule was in line with the needs of the Eastern Bloc scientific community, the actual election was much more controversial, for the Soviet members were supposed to be elected by the assembly at large. In the period between the ad hoc assembly and the general election, the Soviet delegates went to the Soviet Embassy in Copenhagen in order to come up with a common strategy approved by party authorities. It was decided that the Soviet scholars would vote for every other region but refrain from voting for the USSR region—a choice that was to be followed by the participants coming from the other Eastern Bloc countries.[125] The official position of the Soviet Union with respect to the transformation of the ICGRG was communicated through a petition signed by all the Soviet participants: "Due to absence of normal conditions during discussions of various candidates at yesterday meeting (8th July), the Soviet participants of GR6 Conference are abstaining from voting any of the proposed new candidates from USSR. We cannot recognize the results of voting for this Region No. 3 (USSR) and believe that the question of the new Soviet members [of the ICGRG] must be settled by GRG Committee which will be informed of the opinion of the Academy of Science USSR and of the Soviet Gravitational Commission."[126] It is likely, however, that Soviet scientists and those from other Eastern Bloc countries disregarded the official statement and used the secrecy of the vote to express their preference for

[123]"Minutes of the 1st Meetings of a General Assembly towards the foundation of Society for General Relativity and Gravitation GRG," p. 5, DAUT.

[124]Mercier to members of the ICGRG, 30 May 1972, PISGRG, folder 1.2.

[125]Dautcourt, "Bericht über die 6. Internationale Gravitationskonferenz in Kopenhagen," DAUT.

[126]Soviet participants of the GR6 Conference to the Chairman of the Plenary GR6, 9 July 1971, GR6P, Box A-L.

the choice of the new Soviet members, too.[127] The ballot ended with the election of Braginsky and I.D. Novikov—a choice that was considered by the new ICGRG to be final despite the official petition by Soviet participants.

This was the last step in a series of events perceived by Soviet party authorities to be detrimental to Soviet participation in this international organization. The choice of Israel as the host country for the next GR conference, Robinson's vocal attack on Soviet policies, and the transformation of the ICGRG into a democratic society that could elect the Soviet delegates without consulting Soviet scientific institutions were all threats to the participation of Soviet scholars in this international venture within the GRG field.

Immediately after their return, the Soviet ICGRG members were questioned about their inability to protest in a situation that had become politically dramatic for them and about having accepted decisions that they should have opposed. Apparently, there were investigations conducted by party hierarchies on the events in Copenhagen and Fock—who had apparently also rejoined the General Assembly after Braginsky—was punished: he was no longer allowed to go abroad in the future (Khalatnikov 2012, p. 134).[128]

Once Soviet authorities held Fock responsible for his allegedly compliant behavior, and with the uncertain status of the newly elected members, Ivanenko became the Soviet spokesperson for Soviet relations with the ICGRG. In the months following the Copenhagen conference, he attempted to modify what had been decided during the GR6 conference. Firstly, he tried to re-establish the authority of the ICGRG in matters of decisions concerning its membership against the "*street* democracy" of the General Assembly. He argued that there was a "terribly clear difference between the group of all GRG members, elected by GRG (internationally recognized body) possessing good mandates from their respective Academies, Universities, or National Committees etc. and a rather accidental ensemble of a Conference [sic] participants." Since, for instance, participants at this conference comprised 60 Americans versus 11 Soviets, "all allowed in principle to vote," the adopted procedure was clearly unacceptable to Ivanenko.[129] Consequently, he again requested that the election of the Soviet members and the decision made during the General Assembly to replace Wheeler and DeWitt, whose names had not been drawn by Møller's wife, be disregarded. Ivanenko maintained that in order to allow Soviet scientists to participate it was not enough to simply apply national quotas in the composition of the ICGRG. It was necessary for Soviet

[127]The total number of votes that went to the Soviet delegates was the highest (277), together with the number of votes for the American candidates. This is bizarre as one would expect a considerable difference in the case of coordinated abstention, "Minutes of the Meetings on the Committee on GRG, held on occasion of the GR6-Conference in Copenhagen," pp. 18–20, DAUT. Dautcourt confirmed that he did in fact vote for the Soviet candidate. Dautcourt, personal communication.

[128]Apparently, Braginsky and Trautman were also reprimanded and faced difficulties with their superiors. Personal communications by Kip Thorne and Trautman. If we trust Khalatnikov's short description of the events, the report accusing his colleagues was drafted by Petrov.

[129]Ivanenko to Møller, 22 July 1971, GR6P, Box A-L.

central scientific organs to have direct control on the choice of the Soviet members, and this could only be achieved if the ICGRG had a say in the final decision.

The political issue caused by the election of two new Soviet members of the ICGRG was only one of the controversies sparked by the decisions made during the General Assembly, and by far the less traumatic. The ad hoc assembly confirmed that the next international conference would be held in Haifa, Israel, with Nathan Rosen as Chairman of the organizing committee. During the conference in Haifa, the ad hoc assembly agreed, the ISGRG would be established.[130] In fact, this decision jeopardized the chances of scholars from Soviet Bloc countries being able to join the discussions on establishing the new international institutional body. One of the most dramatic potential consequences was that the establishment of an international society on GRG in Israel might have meant the complete exclusion of scholars working in the Soviet Union and in other Eastern Bloc countries.

Mercier and the newly elected President of the ICGRG, Christian Møller, assumed the responsibility of ferrying the community from one structure to another in this stormy sea. Only five days after the conclusion of the GR6 conference, Mercier received an incensed letter from Treder, who had been replaced by his colleague Dautcourt as a member of the ICGRG.[131] Treder protested against the decisions made in Copenhagen. He complained that such important decisions had been made without consulting him and under pressure coming from outside the ICGRG. Treder criticized in particular the decision to transform the committee into a society. In fact, he made it clear that it would not be possible for him nor for his colleagues in the GDR to join this new type of organization. On the other hand, the current structure of the ICGRG suited them very well because it was similar to a private "scholarly club" ("Gelehrten-Klub") aimed at promoting the relativity theory.[132] Official participation in an international society like that envisaged during the Copenhagen conference would consequently have a very different character. Treder used arguments similar to those employed by Ivanenko. As he saw it, the new organization would be more official and then East German scholars would only be able to join as chosen delegates of national institutional bodies, similarly to what happened in the case of international unions such as the IAU and IMU. This implied that Dautcourt, his colleague at ZIAP, would not be able to join the ICGRG, although he had been elected as East German representative of the ICGRG.

After he sent his letter to Mercier, Treder began using all his administrative influence to oblige Dautcourt to retire from the ICGRG. The same day, he wrote an official letter informing Dautcourt that after discussing the matter with the General Secretary of the German Academy of Sciences at Berlin (DAWB)—also

130Mercier to Relativists throughout the World, November 1972, ISGRGR.

131Dautcourt, "Bericht über die 6. Internationale Gravitationskonferenz in Kopenhagen," DAUT; "Minutes of the Meetings on the Committee on GRG, held on occasion of the GR6-Conference in Copenhagen," p. 19, DAUT.

132Treder to Mercier, 15 July 1971, PISGRG, folder 1.3.

responsible for international relations—it was determined that it was not possible for a member of an institute of the academy to join the organization envisaged during the GR6 conference, particularly in view of the decision to establish the new body in Israel. He also ordered Dautcourt to retire immediately from the ICGRG.[133]

Initially, Dautcourt refused, arguing that the Soviet members of the ICGRG had accepted his election and that it had already been decided that all the members of the Eastern Bloc countries would act in unison to avoid the next meeting being held in Israel. Dautcourt also sent a petition to the then President of the DAWB, Hermann Klare, repeating that a withdrawal from the ICGRG—as requested by Treder—would have been "a stab in the back of our Soviet friends."[134] Although Dautcourt declared that he would stand by his decision even if disciplinary measures were taken against him at ZIAP, the pressure was so strong that he had to capitulate. After having tried to maintain a critical position within the ICGRG by supporting the arguments previously presented by Ivanenko and Treder, Dautcourt withdrew in December 1971.[135] The forced retirement of Dautcourt left a vacant position in the ICGRG, which Mercier proposed to give to the East German scholar who got most votes after Dautcourt in the ballot, according to the regional division accepted at the GR6 conference. This was Ernst Schmutzer of the University of Jena.[136] Schmutzer was in fact able to accept and became an ICGRG member, showing that it was possible to circumvent the rule declared by Treder and, under pressure, by Dautcourt that East German members of the transformed ICGRG had to be proposed by the DAWB.[137]

The problems regarding how East German and other Soviet Bloc scholars could join the official international society was only one of the issues under debate. The other was, of course, the chosen location of the following conference. Not only were the diplomatic relations between USSR and Israel interrupted but all Soviet Bloc countries apart from Romania had also suspended diplomatic relations with Israel after the Six-Day War. This situation implied that it would have been impossible for scholars working in these countries to attend the conference. This was not the best way to launch a new international society, which was itself problematic for these scholars who were not going to attend the meeting.

---

[133]Treder to Dautcourt, 15 July 1971, DAUT.

[134]Dautcourt to Hermann Klare, 20 October 1971, DAUT.

[135]Dautcourt to Møller 9 December 1971; Dautcourt to Møller, 10 December 1971, DAUT; and Dautcourt to Møller, 29 October 1971, PISGRG, folder 1.3.

[136]Mercier to the members of the ICGRG, 30 May 1972, PISGRG, folder 1.2. Although Bergmann had nothing against Schmutzer, he wanted to better understand the motivations behind Dautcourt's withdrawal. He did not buy Mercier's explanation that Dautcourt disliked the idea that he was "taking the place so brilliantly occupied by his former teacher [sic], H. J. Treder, precisely at the time when […] the latter has been very ill for a prolonged period." Mercier to the members of the ICGRG, 30 May 1972, PISGRG, folder 1.2; and Bergmann to Mercier, 13 June 1972, PISGRG, folder 1.3.

[137]Schmutzer to Mercier, 24 April 1973, PISGRG, folder 1.3.

Ivanenko saw the modification of this plan as a necessary step to preserve the ICGRG and the useful work of unification it had carried out since 1959, a position also held by other Soviet Bloc members of the ICGRG.[138] In response to Ivanenko's requests, Møller replied amiably, declaring the best of intentions to work together "on the task to keep the unity of relativists in all parts of the world." To do so, he proposed to use the reciprocal influence to avoid hasty action from scholars working on both sides of the Iron Curtain, such as the "disturbing" letter received from Treder and Robinson's "foolish action." As for the specific points raised by Ivanenko, Møller asked him to accept the decision of the new Soviet and American members, while working together to restore a decisional role of the centralized Soviet institutions for the election of the next ICGRG members.[139]

As Møller avoided mentioning the sensitive issue of Israel in his reply, Ivanenko changed his strategy. The Italian physicist Bruno Bertotti, who had replaced Cattaneo as ICGRG member during the GR6 conference, made the proposal of hosting the international conference at the Ettore Majorana Centre for Scientific Culture in the medieval town of Erice in Sicily. According to Bertotti, who was apparently unaware that the decision to hold the conference in Israel was final, "the smallness and the beauty of Erice [would] favor the scientific contact and relaxed atmosphere at the meeting," surely more than Haifa, where a section of the community would not go due to the lack of diplomatic relations between their countries and Israel.[140] Ivanenko and other scholars working on the Eastern side of the Iron Curtain immediately and enthusiastically supported Bertotti's proposal.[141]

The stage was set for painstaking negotiations involving both the issues concerning the structure of the international organization and where it would be established. The "distressing [...] news [...] that the Soviets [were] making difficulty with the choice of Israel," prompted Western Bloc scholars to reflect on the political implications of the choice, while they continued to maintain the view that the world of science should "reject political issues."[142] However, as British mathematician Clive Kilmister recognized, it was very difficult "to take any decisions which [were] free of political pressures in one direction or the other."[143] As the majority of scholars working on the Western side of the Iron Curtain thought

[138]Ivanenko to Møller, 22 July 1971, GR6P, Box A-L. During the dramatic period in which he was trying to maintain his membership despite the pressure from Treder, Dautcourt argued that "[t]he exposed situation and its one-sided orientated policy make Israel appear as unsuitable for organizing an international conference" -"[d]ie exponierte Lage und seine einseitig orientirte Politik lassen Israel als ungeeignet für die Ausrichtung einer internationalen Konferenz erscheinen." Dautcourt to Møller, 29 October 1971, PISGRG, folder 1.3.

[139]Møller to Ivanenko, 10 September 1971, GR6P, Box A-L.

[140]Bertotti to Mercier, 21 September 1971, PISGRG, folder 1.3.

[141]Dautcourt to Møller, 10 December 1971, PISGRG, folder 1.3; Ivanenko to Mercier, undated, probably June 1972, attached to Mercier to sub-committee, 14 July 72, PISGRG, folder 1.3; Wheeler to Møller, 4 April 1972, JWP, Box 18, folder Møller.

[142]Wheeler to Møller, 4 April 1972, JWP, Box 18, folder Møller.

[143]Kilmister to Mercier, 12 June 1972, PISGRG, folder 1.3.

that all the political "pressures to move the conference away from Israel must be resisted,"[144] the decision to hold the GR7 conference in Haifa was publicly restated again and again, and was finally supported by the official confirmation that IUPAP had nothing against the decision to organize the conference in Israel.[145]

Ivanenko protested once again against the decision of Israel hosting the GR7 with an official letter that was circulated among the members of the subcommittee responsible for the statute explaining why, in his view, it was not possible to accept this venue: "The question is not only of participating at scientific discussions, but a very important point concerns taking part at the scheduled meetings of the GRG-Committee where very important questions of the statute, election of the new chairman, of the new members etc. must be settled. I confess I cannot imagine that all this could be reasonably well proceed without participation of representatives of the Soviet Union and probably also some other countries." As a possible diplomatic solution, Ivanenko proposed organizing a conference in Italy either *before* or *after* the general conference in Israel.[146] This proposal was then made official by Bertotti who offered to organize a conference called "Experiments on Gravitation" three days before the GR7 conference, clarifying that it should "in no way interfere with the General Conference in Israel."[147]

The proposal of organizing two different conferences was taken into serious consideration and discussed at the meeting of the subcommittee on the statute, which met in Geneva on 20 September 1972. The participants attempted to find a way out of the political impasse by recommending "that the establishment (different from foundation) of the new international Society on GRG be practically realized at a meeting which could be held either before or after the international Conference GR7 at some suitable place to be decided upon by the International Committee on GRG."[148] Mercier, Møller, and the members of the subcommittee were trying to mediate with a solution that was acceptable for all the parties involved. Bertotti's proposal to host a conference in Italy before or after the conference in Haifa was, however, difficult to accept because it was impossible for this conference not to interfere with the official GR7.[149] On the other hand, it was clear to some members of the ICGRG that "the official foundation of the new society [could not] take place in Israel if [their] colleagues from Eastern countries for reasons beyond our control [were] absent."[150] Møller's final proposal was then to establish the society at

---

[144]Thorne to Mercier, 25 July 1972, PISGRG, folder 1.3.

[145]Mercier to the members of the ICGRG, 30 May 1972, PISGRG, folder 1.2; Mercier to Rosen, 10 October 1972, PISGRG, folder 1.3.

[146]Ivanenko to Mercier, undated, probably June 1972, attached to the letter from Mercier to sub-committee, 14 July 72, PISGRG, folder 1.3.

[147]Bertotti to Møller, 12 September 1972, PISGRG, folder 1.3.

[148]Mercier to Rosen, 10 October 1972, PISGRG, folder 1.3.

[149]Møller to Bertotti, 18 October 1972; and Rosen to Mercier, 9 January 1973, PISGRG, folder 1.3.

[150]Møller to Bertotti, 18 October 1972, PISGRG, folder 1.3.

another general conference, which could possibly be held in Italy one year after the GR7.

Amidst the heated negotiations on where to establish the society, the subcommittee on the statute elected at the GR6 ad hoc General Assembly was in the process of writing a new draft of the statute. One central theme was the conflict between individual membership and national delegations. Kip Thorne openly expressed this conflict and his position on this issue in a letter to the member of the subcommittee Dennis Sciama: "[T]o be viable, our organization must be composed of individual scientists and not of national delegations. On the other hand, the Soviet bureaucracy insists on regarding all such organizations as composed of national delegations. Somehow, a constitution must be drafted which makes it perfectly clear that ours is an organization of individual scientists; but the phraseology should probably be such that Soviet bureaucrats can misinterpret it if they wish."[151]

The final proposal made by the subcommittee was to permit two different kinds of membership: individual membership and corporate membership, which, according to Mercier, met the wishes expressed by Thorne in his letter to Sciama.[152] This regulation was the most explicit result of the negotiations between the different needs expressed by scientists working in different political systems during the Cold War. Through this article, the authors were institutionalizing the political character of the enterprise by giving the society envisioned a hybrid character—a feature that was almost unique in contrast to the array of international scientific institutions existing during this period (see Chap. 3).[153] When Mercier sent the draft constitution to the members of the ICGRG, he underlined the necessity of taking the right political actions favoring the creation of a truly international body that would not exclude any of the experts working in the field. These actions were even more necessary in the light of the choice of where to establish the society, for which Mercier made an explicit personal appeal asking the ICGRG members to show "a spirit of generosity" toward different needs and points of view.[154]

As decided by the ad hoc assembly at the Copenhagen meeting, Mercier circulated the statute among all the scientists on his list of experts, addressing the letter to "Relativists throughout the World." He requested the "relativists" to provide

---

[151]Thorne to Sciama, 1 November 1972, PISGRG, folder 1.3.

[152]Draft Constitution of The International Society for General Relativity and Gravitation, attached to the letter from Mercier to the members of the ICGRG, 18 November 1972, PISGRG, folder 1.2.

[153]Among the international scientific institutional bodies of that period, the only two that seemed to resemble the structure of the ISGRG were the International Society of Biometeorology, founded in 1956 and the International Society of Electrochemistry established in 1970 as the evolution of the *Comité International de Thermodynamique et de la Cinétique Electrochimiques,* founded in 1949 (see Tromp 1960; Bockris 1991; Tannenberger 2000). I am grateful to Helmut Tannenberger and Dieter Landolt for information concerning the structure and historical developments of the CITCE-ISE.

[154]Mercier to the members of the ICGRG, 18 November 1972, PISGRG, folder 1.2.

comments on the statute in order to finalize it during the next meeting of the ICGRG, which would be held in Paris in June 1973.[155] Through the creation of a hybrid society to which both individuals and institutional bodies could adhere, the members of the subcommittee hoped that they had solved all the major political issues and that all the interested scholars could accept the statute independently of the political context in which they lived. This hope was misplaced. Eminent members of both American and Soviet GRG communities responded negatively. The Soviets pushed to maintain the existing nature of the committee, while the Americans urged to make the society more open and democratic.[156] Bergmann and Goldberg even informed Mercier that they were collecting the comments of American scholars to prepare a complete revision of the statute.[157] In the meantime, Mercier received informal news from Rosen that many American scientists were against the decision to hold a large meeting in Europe to establish the society one year after the GR7 conference for logistic, economic, and political reasons. According to Rosen, the American physicists he met at the 1972 Texas Symposium of Relativistic Astrophysics still believed that the ISGRG should be established in Israel as had been agreed during the Copenhagen conference.[158]

Mercier was incensed by all the replies that he understood as open opposition to his attempts to find reasonable diplomatic solutions to what were evidently political problems in the path to creating an international institutional body—and a quite original one—in the Cold War context. He felt that his actions and attempts were attacked from all sides, as he wrote to Bergmann and Goldberg: "The baby called International Society on GRG is far from being born yet. Not only you, by Sovjet [sic] Colleagues and others seem to prepare the attack against the draft. Although we have done all we could in order to have a structure which would allow for everybody to join. I am not sure that under these circumstances we shall come to an understanding very soon."[159] He replied with the same tone to the Soviet colleagues by stressing that the constitution was "very favourable to Sovjet [sic] Scientists," in that it made "easy for [them] to collaborate through their belonging to official Institutions which could become corporate members."[160] To Rosen, he once more showed his antipathy for the attempt of those gathering at the Texas symposia to influence the activity of the ICGRG and the GRG international community: "I am annoyed by the fact the you have postponed any practical step *by consulting again people who do not really represent the bulk of all relativists* and putting new

155André Mercier to Relativists throughout the World, November 1972, ISGRGR.

156A. A. Sokolov to Mercier, 10 January 1973, PISGRG, folder 1.2; N. V. Mitskiévic to Mercier, 11 January 1973, PISGRG, folder 1.2; Bergmann and Goldberg to Mercier, 25 January 1973, PISGRG, folder 1.2.

157Bergmann and Goldberg to Mercier, 25 January 1973, PISGRG, folder 1.2.

158Rosen to Mercier, 9 January 1973, PISGRG, folder 1.3.

159Mercier to Bergmann and Goldberg, 1 February 1973, PISGRG, folder 1.2.

160Mercier to Mitskiévic and Sokolov, 1 February 1973, PISGRG, folder 1.2.

suggestions and solving no problem [...] Sometimes, I ask myself if it is does not make any sense to spend such a great lot of my time to get nowhere."[161]

Under the impression that the compromise he, Møller, and the members of the subcommittee on the statute had prepared was failing to obtain the necessary support from all the parties involved, Mercier began thinking that the ISGRG would "never be born."[162] Because of the "absurd stress" this situation was placing on the GRG community and, without doubt, on him personally, Mercier decided to avoid any further action before a final decision that was to be made by the President and the members of the ICGRG during its next meeting in Paris. However, he did not leave the discussion without repeating his ideological message that to solve the problem was an act of "peace" which implied a full understanding of each other position—"a matter of love," as he declared in one of his most passionate letters.[163]

In the months following what Mercier saw as the lowest point in the attempts to build a solid international institutional setting for research in the field of GRG, things began to settle down. Møller was much less pessimistic than Mercier as, in his view, the majority of "relativists" were eager to continue the international collaboration. Since the GR7 would be held in Israel without any question and the Soviet colleagues would not attend "for reasons beyond our control," he suggested supporting the proposal to found the society by mail as the only practical possibility. If the majority of ICGRG members agreed with this decision during the next meeting in Paris, Møller proposed to proceed this way, asking members to refrain from suggesting other "petty changes" to the proposed statute. Once 150 written consents from at least ten different nations arrived by mail, the ISGRG could be declared as established. This procedure would allow the ISGRG to be founded before the GR7 in Israel. As for the membership of scholars from Eastern Bloc countries, Møller again professed his optimism: "It might be difficult for our Eastern colleagues to commit themselves in the first stage but as soon as the Society exists I am sure that they will find a way to join, if not as individuals then through their Academies."[164]

The revised draft of the constitution prepared by Bergmann and Goldberg following the suggestions of many American scholars did not contain any radical departure from the nature of the proposed International Society and adhered to what had been agreed on the most sensitive issues: the option of different kinds of memberships—both individual and corporate—and the name of the official governing body of the society, which remained the ICGRG.[165] The ICGRG was

---

[161]Mercier to Rosen, 18 January 193, PISGRG, folder 1.3, emphasis mine.

[162]Mercier to Utiyama, 1 February 1973, PISGRG, folder 1.3.

[163]Mercier to Rosen, 1 February 1973, PISGRG, folder 1.3. In this letter, Mercier declared that he understood Rosen's point of view, but he was worried that no feasible solution had been proposed. He was trying to define the "philosophical condition of peace," in agreement with the views he had already expressed in his writings (Mercier 1959).

[164]Møller to Mercier, 5 February 1973, CMP, folder GR7 1974.

[165]Bergmann and Goldberg, Memorandum: Revised Draft of GRG Constitution, 28 March 1973, RISGRG.

relieved to accept this version as the constitution that would be proposed to the entire GRG community.[166] The issue of where the society would be established was finally solved by accepting Møller's proposal.[167] After initial resistance, Mercier eventually agreed to the suggestion advanced by Rosen and others that voting by mail was the easiest way to find a plausible practical solution to a complex political issue.[168]

Mercier sent the final constitution with the application forms to his list of scientists in October 1973. Less than three months later, on 7 January 1974, Mercier informed Møller that as many as 166 membership fees, including both individual and corporate, had been paid from more than 23 nations. The ISGRG could then "be declared into existence."[169] Among the members, there were three individuals from the Soviet Union whose membership fees were paid by the Academy of Sciences of the USSR.[170] At the time the ISGRG was established, these were the only representatives of the entire Soviet Bloc community. Despite the limited number of Eastern Bloc members in the new society, the three membership fees constituted a clear sign that Soviet scientific and political authorities had approved the ISGRG. Neutral Switzerland would officially continue to be the geographical basis of the new international society as the statute was drafted according to Swiss legislation. At the beginning of 1974, the long path toward the institutionalization of the field of GRG in the international panorama had come to a conclusion. It was a difficult journey that required a long series of negotiations, compromises, and discussions. The approved statute of the ISGRG contained the codification of this controversial history: its final form was a hybrid society that both Western and Eastern Bloc scholars could join and, if elected as members of the ICGRG, administer.[171]

Yet, while the decision to establish the society via mail resolved the political crisis concerning Israel, it did not fix the problems concerning how to choose the governing body of the ISGRG, which was still called ICGRG. Since the General Assembly could not elect the Nominating Committee—as the General Assembly would first take place at the GR7 conference in Israel—Mercier invited Møller to appoint an ad hoc Nominating Committee that reflected the geographical and political distribution of the GRG community. Mercier again provided the list of countries and geographical areas that should have been represented in the Nominating Committee: the U.S., USSR, U.K. or Commonwealth (excluding Asia),

---

[166]Mercier to Y. Choquet-Bruhat and D. W. Sciama, 21 May 1973, PISGRG, folder 1.2.

[167]Mercier to Relativists throughout the World, 10 October 1973, PISGRG, folder 1.4.

[168]Mercier to Rosen, 1 February 1973; and Rosen to Mercier, 23 February 1973, PISGRG, folder 1.3; Møller to Mercier, 5 February 1973, CMP, folder GR7 1974.

[169]Mercier to Christian Møller, 7 January 1974, PISGRG, folder 1.4.

[170]Mercier to Members at large of the International Society on GRG, 28 March 1974, ISGRGR. The three Soviet members were Vladimir A. Belinsky, Khalatnikov, and Evgeny Lifshitz, all working in institutes of the Soviet Academy of Sciences in Moscow.

[171]Constitution of the International Society for General Relativity and Gravitation (as finalized by the ICGRG at its meeting in Paris on 22 June 1973), ISGRGR.

France and other Latin European countries, German-speaking countries and Eastern Europe, Asia, and other "places." Mercier also took the liberty of suggesting a list of those scholars who, in his opinion, might be chosen by Møller as members of the ad hoc Nominating Committee. In his precise instructions to Møller, Mercier included diplomatic reasoning that was not an integral part of the statute. Mercier clearly expressed the need for the future ICGRG to maintain the symmetry "between USA and USSR," implying that the two nations should continue to have the same number of members in the governing body of the society. Although the officials would be elected later by the General Assembly, Mercier was conscious that implicit steps had to be taken to maintain the balance between the United States and the Soviet Union through the carefully considered selection of the members of the Nominating Committee.[172] On the other hand, Mercier continued to propose a geographical area that included all German-speaking countries, rather than clearly dividing the European countries according to the political spheres of influence. In his duty as non-voting Secretary of the Nominating Committee, Mercier continued to give his opinions on how the vote should be carried out. He also continued to influence Møller by providing suggestions about possible names that could be included in the list of nominees. Thus, Mercier had an impact on many of Møller's decisions, as the President of the ICGRG was counting on Mercier to have the list of the nominees in time for the election during the General Assembly at the GR7 conference.[173]

Moreover, because of the limited amount of time, Mercier had to make decisions on summarizing the votes of the members of the Nominating Committee, which significantly affected the final list of nominees.[174] In the events that led to the first democratic election of the governing body of the ISGRG, it seemed that Mercier played quite a relevant role although in theory he was not supposed to have any active part in the decisions of the Nominating Committee. All the members of the Nominating Committee, as well as the ISGRG at large, recognized his enormous involvement in the organization and his decisive role in its developments. None of the members of the Nominating Committee expressed any opposition to the re-election of Mercier as Secretary of the ISGRG, which Møller deemed "imperative."[175]

On 24 June 1974, the first meeting of the General Assembly of the newly established ISGRG took place in the Mexico Building of Tel-Aviv University. (For logistic reasons, the GR7 conference was eventually held in Tel-Aviv instead of Haifa).[176] The 110 attendees were able to learn about the status of the new ISGRG from Mercier and could vote for the replacement of eight members of the ICGRG. Despite the absence of participants from the Soviet Bloc, the first General Assembly

---

[172] Mercier to Møller, 3 November 1973, PISGRG, folder 1.4.

[173] Mercier to Møller, 1 December 1973, PISGRG, folder 1.4.

[174] Mercier to the Members of Nominating Committee GRG, 27 March 1974, PISGRG, folder 1.4.

[175] Møller to Mercier, 10 April 1974, PISGRG, folder 1.4.

[176] Rosen to Møller, 26 March 1973, PISGRG, folder 1.2.

showed that the steps undertaken had been successful.[177] Only one modification to these actions was requested: a neater political separation in the Nominating Committee. Trautman restated the principle proposed by Wheeler to substitute the purely geographical/linguistic/cultural partition preferred by Mercier with a political division. Trautman proposed that the Soviet member of the Nominating Committee be the representative of East Germany and the other Eastern Bloc nations as well, while the representative of West Germany should be asked to represent West Germany, Austria, and Switzerland.[178]

Besides this change, the new society did not suffer from any other issues due to political considerations. The difficult negotiations had led to the establishment of a scientific institution that had almost no parallel in the middle of the Cold War. The year after the GR7 conference, the status of the ISGRG became even more official at the international level, when it formally became the second Affiliated Commission of IUPAP (Anon. 1992, p. 37). Mercier could finally be relieved that the enormous efforts by himself and others had eventually led to the establishment of an international society that included the participation of scholars from throughout the world. The ISGRG in fact became one of the first instances of an international scientific association not based on the rule of national membership in which members of both Western and Eastern Bloc countries participated and could be democratically elected in its governing body by all the members.

References

Anon. 1965. 50 Jahre Allgemeine Relativitätstheorie—60 Jahre Spezielle Relativitätstheorie. *Bulletin on General Relativity and Gravitation* 9: 4–5. doi:10.1007/BF02938033.

Anon. 1967. Short statement. *Bulletin on General Relativity and Gravitation* 14: 11. doi:10.1007/BF02938018.

Anon. 1992. *UIPPA-IUPAP 1922–1992*. Album souvenir realized in Quebec by the Secretariat of IUPAP. http://iupap.org/wp-content/uploads/2013/04/history.pdf. Accessed 7 March 2016.

Bergmann, Peter. 1962. Review of Infeld Festschrift. Recent developments in general relativity. *Science* 138: 134.

Bergmann, Peter. 1963. Review of Gravitation: An introduction to current research, ed. Louis Witten. *Science* 140: 654. doi: 10.1126/science.140.3567.654-a.

Blum, Alexander. 2016. The conversion of John Wheeler. Talk Presented at the 7th International Conference of the European Society for the History of Science. Prague, 22 September 2016.

Bockris, J. O'M. 1991. The founding of the international society for electrochemistry. *Electrochimica Acta* 36: 1–4. doi:10.1016/0013-4686(91)85171-3.

Collins, Harry. 2004. *Gravity's shadow: The search for gravitational waves*. Chicago: University of Chicago Press.

Dicke, R.H., P.J.E. Peebles, P.G. Roll, and D.T. Wilkinson. 1965. Cosmic black-body radiation. *The Astrophysical Journal* 142: 414–419. doi:10.1086/148306.

[177]Minutes of the General Assembly of the International Society on GRG, 25 June 1974, ISGRGR.

[178]Andrzej Trautman to Mercier, 7 May 1975, PISGRG, folder 1.4.

Elzinga, Aant. 1996. Modes of internationalism. In *Internationalism and Science*, ed. Aant Elzinga, and Catharina Landstrom, 3–20. London: Taylor Graham.

Fennell, Roger. 1994. *History of IUPAC, 1919-1987*. Oxford: Blackwell Science Ltd.

Fock, Vladimir et al. 1968. *GR5, abstracts of the 5th international conference on gravitation and the theory of relativity*. Tbilisi: Publishing House of the Tbilisi University.

Franklin, Allan. 1994. How to avoid the experimenters' regress. *Studies in History and Philosophy of Science Part A* 25: 463–491. doi:10.1016/0039-3681(94)90062-0.

Franklin, Allan, and Harry Collins. 2016. Two kinds of case study and a new agreement. In *The philosophy of historical case studies*, ed. Tilman Suaer, and Raphael School, 95–121. Charm: Springer. doi:10.1007/978-3-319-30229-4_6.

Gieryn, Thomas F. 1999. *Cultural boundaries of science: Credibility on the line*. Chicago: University of Chicago Press.

Golan, Galia. 1990. *Soviet policies in the Middle East: From World War Two to Gorbachev*. Cambridge: Cambridge University Press.

Govrin, Yosef. 1998. *Israeli-Soviet relations, 1953–67: From confrontation to disruption*. London: OR Frank Cass.

Greenaway, Frank. 1996. *Science international: A history of the International Council of Scientific Unions*. Cambridge: Cambridge University Press.

Hawking, Stephen W, and George F.R. Ellis. 1973. *The large scale structure of space-time*. Cambridge: Cambridge University Press.

Held, Alan, Heinrich Leutwyler, and Peter G. Bergmann. 1978. To André Mercier on the occasion of his retirement. *General Relativity and Gravitation* 9: 759–762. doi: 10.1007/BF00760862.

Hentschel, Klaus, and Monika Renneberg. 1995. Eine akademische Karriere: Der Astronom Otto Heckmann im Dritten Reich. *Vierteljahrshefte für Zeitgeschichte* 43: 581–610.

Hoffmann, Dieter. 2013. Fifty years of physica status solidi in historical perspective. *Physica Status Solidi (b)* 250: 871–887. doi:10.1002/pssb.201340126.

Hoffmann, Dieter. 2017. In den Fußstapfen von Einstein: Der Physiker Achilles Papapetrou in Ost-Berlin. In *Deutsch-griechische Beziehungen im ostdeutschen Staatssozialismus, 1949–1989*, ed. Konstantinous Kosmas. Berlin: Romiosini. (in print).

Ivanenko, Dimitri. 1969. Report on the Ahmedabad conference 1969. *Bulletin on General Relativity and Gravitation* 20: 1. doi:10.1007/BF02907867.

Khalatnikov, Isaak M. 2012. *From the atomic bomb to the Landau Institute: Autobiography. Top Non-Secret*. Berlin, Heidelberg: Springer Berlin Heidelberg.

Kilmister, Clive W. 1963. Review of Gravitation: An introduction to current research. Ed. Louis Witten. *Planetary and Space Science* 11: 997. doi: 10.1016/0032-0633(63)90130-2.

Kragh, Helge. 1996. *Cosmology and controversy: The historical development of two theories of the universe*. Princeton, NJ: Princeton University Press.

Laitko, Hubert. 1999. The reform package of the 1960s: The policy finale of the Ulbricht Era. In *Science Under Socialism: East Germany in Comparative Perspective*, ed. Kristie Macrakis, and Dieter Hoffmann, 44–63. Cambridge, MA: Harvard University Press.

Laron, Guy. 2010. Playing with fire: The Soviet–Syrian–Israeli triangle, 1965–1967. *Cold War History* 10: 163–184. doi:10.1080/14682740902871869.

Lehmkuhl, Dennis. 2017. The discovery of gravitational waves: Narrative and facts. Talk presented at the workshop *Political Epistemology: New Approaches, Methods and Topics in the History of Science—II*, Berlin, March 23 2017.

Little, Douglas. 2010. The Cold War in the Middle East: Suez crisis to Camp David accords. In *The Camrbidge History of the Cold War, Vol. II, Crises and Détente*, ed. Melvyn P. Leffler, and Odd Arne Westad, 305–326. Cambridge: Cambridge University Press.

Longair, Malcolm S. 2006. *The cosmic century: A history of astrophysics and cosmology*. Cambridge: Cambridge University Press.

Mercier, André. 1959. *De l'amour et de l'etre*. Paris: Louvain.

Mercier, André. 1965. Report on the Second Texas Symposium on relativistic astrophysics. *Bulletin on General Relativity and Gravitation* 8: 1–12. doi:10.1007/BF02938015.

Mercier, André. 1966. Golden jubilee celebrations of the publication by Einstein of the theory of general relativity and gravitation. *Bulletin on General Relativity and Gravitation* 10: 1–10. doi:10.1007/BF02938021.

Mercier, André. 1967. Report on the Third Texas Symposium—1967. *Bulletin on General Relativity and Gravitation* 14: 1–10. doi: 10.1007/BF02938017.

Mercier, André. 1979. Birth and Rôle of the GRG-organization and the cultivation of international relations among scientists in the field. In *Albert Einstein: His influence on Physics, Philosophy and Politics*, ed. Peter C. Aichelburg, and Roman U. Sexl, 177–188. Braunschweig: Vieweg. doi: 10.1007/978-3-322-91080-6_13.

Peebles, Phillip James Edwin. 2017. Robert Dicke and the naissance of experimental gravity physics, 1957–1967. *The European Physical Journal H* 42: 177–259. doi:10.1140/epjh/e2016-70034-0.

Penrose, Roger. 1965. Gravitational collapse and space-time singularities. *Physical Review Letters* 14: 57–59. doi:10.1103/PhysRevLett.14.57.

Recent developments in general relativity. 1962. Oxford, New York: Pergamon Press.

Rickles, Dean. 2011. The Chapel Hill Conference in context. In *The Role of Gravitation in Physics: Report from the 1957 Chapel Hill Conference*, ed. Cécile DeWitt-Morette, and Dean Rickles, 7–21. Berlin: Edition Open Access.

Ro'i, Yaacov, and Boris Morozov. 2008. *The Soviet Union and the June 1967 Six Day War*. Stanford: Stanford University Press.

Ruffini, Remo. 2010. Moments with Yakov Borisovich Zeldovich. In *The Sun, the Stars, The Universe and General Relativity: International Conference in Honor of Ya.B. Zeldovich's 95th Anniversary, AIP Conference Proceedings* 1205: 1–10. doi: 10.1063/1.3382329.

Savranskaya, Svetlana, and William Taubman. 2010. Soviet Foreign Policy, 1962–1975. In *The Cambridge History of Cold War, Vol. II, Crises and Détente*, ed. Melvyn P. Leffler, and Odd Arne Westad, 134–157. Cambridge: Cambridge University Press.

Schwab, George (ed.). 1981. *Eurocommunism, the ideological and political-theoretical foundations*. Westport: Greenwood Press.

Shlaim, Avi. 2014. *The iron wall: Israel and the Arab world*. New York: WWNorton & Company.

Stolarik, M. Mark (ed.). 2010. *The Prague Spring and the Warsaw Pact invasion of Czechoslovakia, 1968: Forty years later*. Mundelein, Ill.: Bolchazy-Carducci Publishers.

Tannenberg, Helmut. 2000. From CITCE to ISE. *Electrochimica Acta* 45: xxvii–xxviii.

Thorne, Kip S. 1994. *Black holes and time warps: Einstein's outrageous legacy*. New York: WWNorton & Company.

Treder, Jürgen (ed.). 1966. *Entstehung, Entwicklung und Perspektiven der Einsteinschen Gravitationstheorie*. Berlin: Akademie Verlag.

Trimble, Virginia. 2017. Wired by Weber. *The European Physical Journal H* 42: 261–291. doi:10.1140/epjh/e2016-70060-5.

Tromp, Solco. 1960. Report of the secretary—treasurer. *International Journal of Bioclimatology and Biometeorology* 4: 213–214.

Weart, Spencer. 1992. The solid community. In *Out of the crystal maze: Chapters from the history of solid state physics*, ed. Lillian Hoddeson, Ernst Braun, Jurgen Teichmann, and Spencer Weart, 617–669. Oxford: Oxford University Press.

Wheeler, John A. 1968. Our universe: The known and the unknown. *The American Scholar* 37: 248–274.

Will, Clifford. 1986. *Was Einstein right?: Putting general relativity to the test*. New York: Basic Books.

Will, Clifford. 1989. The renaissance of general relativity. In *The new physics*, ed. Paul Davies, 7–33. Cambridge: Cambridge University Press.

Wilson, Benjamin, and David Kaiser. 2014. Calculating times: Radar, ballistic missiles, and Einstein's relativity. In *Science and technology in the global cold war*, ed. Naomi Oreskes, and John Krige, 273–316. Cambridge, MA: The MIT Press.

Witten, Louis (ed.). 1962. *Gravitation: An introduction to current research*. New York: Wiley.

Chapter 6
Conclusion

Abstract In the conclusion, it is argued that the institutional history of general relativity and the final structure of the International Society on General Relativity and Gravitation were somewhat peculiar, if not unique, with respect to the various institutionalization processes of international scientific cooperation in the post-World War II period. After reviewing the various cultural, social, political, and epistemic tensions characterizing the development of the international community and its institutional representation, possible explanations of why the field of general relativity was conducive to the establishment of this quite original institutional structure are addressed. It is argued that this question cannot be separated from questions concerning the effective relevance of the community-building activities for the process of the renaissance of general relativity. These activities shaped the return of general relativity to the mainstream of physics during the formative phase of the "General Relativity and Gravitation" community. Other relevant factors were the special character of general relativity—which was perceived to be unrelated to any direct military applications—and the role model Einstein embodied for scientists working on both sides of the Iron Curtain. Finally, a programmatic reflection on the function held by scientists/community builders in the evolution of scientific knowledge is proposed, based on André Mercier's role in this specific case.

Keywords Albert Einstein · André Mercier · Cold War · Community building · General relativity · Internationalism in science · International Committee on General Relativity and Gravitation · Relativistic astrophysics · Scientific institution

From the first tentative attempts to strengthen the contacts between the few research groups working in areas related to general relativity in the early 1950s to the formation of an international society dedicated to the field of GRG in the mid-1970s, the community-building activities in the GRG domain had come a long way. As had the scientific field itself. General relativity was no longer considered a marginal field—closer to mathematics than physics—and by the mid-1970s was seen as an exciting branch of theoretical physics with many links with experimental

R. Lalli, *Building the General Relativity and Gravitation Community During the Cold War*, SpringerBriefs in History of Science and Technology,
DOI 10.1007/978-3-319-54654-4_6

and observational programs. It is tempting to see the connections between these social and epistemic processes as unproblematic by assuming that the progress of the field simply led to establishing and developing its own institutional configuration. As I argued in the previous chapters, however, this was not the case.

The formation of the institutional body that promoted the emerging field of GRG was mired in tensions of a surprisingly varied nature. Even in the early years, the explicit community-building activities were shaped by regional and cultural divisions—particularly related to differences in the European and the American approach both to the field and to scientific exchange—symbolized by the opposing styles of the Bern and Chapel Hill conferences. Once the ICGRG was established, this tension was still present in the self-identification of the institutional body through the debate on which should be considered the first GR conference in the series. The debate was resolved by Mercier who conceded the status of first international conference to the Chapel Hill conference but suggested the title of GR0 for the Bern conference. In other words, the Bern conference maintained the official status of the initiator of the new "relativistic" era.

Once the ICGRG was established, a more dramatic tension emerged: the conflict between the authoritative scientists who were ruling the recently established institutional body and a rapidly growing younger generation of scholars who contested the authority of this self-appointed group. This clash between different generations was also a manifestation of the different regional cultures of scientific practice, as the American, or, rather, Anglo-American, groups were seemingly the most vocal in their demand for change. This tension then could not be simply reduced to a generational conflict related to the struggle for power. It also entailed deeper conflicts concerning different visions of the priorities of the scientific field the ICGRG was built to implement. Furthermore, the generational tension was articulated as a cultural conflict about what the social dimension of science should be: a closed elitist scientific group versus an open democratic society.

Another kind of tension, which cannot be separated from those mentioned above, was of an epistemic nature. The epistemic status of the field was very much uncertain and openly debated. Specifically, the origin of the tension was the question of which macro-discipline general relativity belonged to, or, to put it another way, the issue of the preferential connections between general relativity and the disciplines of physics, mathematics, and astronomy. Since entering the field, Wheeler had held and defended very clear views on how general relativity should be restored to where it belonged, namely, to pure physics. Not all members of the ICGRG agreed with Wheeler's stance, and those who agreed in principle—as Mercier did—held quite different opinions about how these links with physics should be developed in practice. Mercier's approach, for instance, would have been to move in the direction of training a new generation of theorists with a much stronger philosophical baggage. The focus on general relativity, in Mercier's view, should have helped theoretical physicists ask deeper foundational questions when constructing theories, which could have then led to major advances. For Wheeler, on the other hand, the return to physics would have been through drawing precise links with the physical world. The serendipitous discovery of quasars in 1963 offered, for the first time, the opportunity to explore the links between theory in GRG and observable physical objects of a completely new nature. As it happened, this immediately led to formation of a brand new research field: relativistic astrophysics. This discovery and its tremendous implications were in line with Wheeler's program but also became a source of tension within the ICGRG when the committee attempted to include the exploding field of relativistic astrophysics within the structure that was already being built.

The last and most evident kind of tension was political. In the polarized political context of the Cold War, it was impossible to avoid the intermingling of scientific and political matters in the pursuit to establish an international institutional body. The very possibility of establishing the committee was a result of changing political contexts: the post-Stalinist détente meant more relaxed East-West relations—which in turn allowed the opening of an epoch of international scientific collaboration. And with its neutral policies, Switzerland was able to serve as the meeting place for scientists working on opposite sides of the Iron Curtain, also in the case of the GRG community. The ICGRG came to embody the ideal of pure scientific enterprise and scientific internationalism, raising the hopes of some of its members that the ICGRG might be a step, albeit a small one, toward promoting peaceful relations. Political considerations certainly played a role in shaping the structure of the ICGRG, although the participants tended to consider the decisions they made simply as a response to scientific necessities. By mimicking the functioning of more established forms of international collaboration—particularly concerning the implicit rules of a kind of national/regional representativeness within the membership and the related balance between the number of American and Soviet members—the ICGRG was able to work quite well within the Cold Was context. The ICGRG members were able, in fact, to overcome all possibly political conflicts right up until 1967. This pattern was broken, first by the Six-Day War, which led to the rupture of diplomatic relations between the Soviet Union and Israel, and, one year later, by the Czech crisis with all its political consequences, particularly in Europe.

These armed conflicts impacted the lives of some of the members of the GRG community or some of their close acquaintances. Politics then entered the work of the ICGRG and its activity in a very divisive manner. This state of affairs led the actors to perform boundary-work by attempting to draw a clear line between science and politics. These actors tried to define what were unacceptable political interventions within what they wanted to maintain as a purely scientific activity. But different scientists held quite different views about what was admissible or inadmissible political discourse. For Fock, holding a conference in Israel in 1974 in that particular situation in the Middle East was a political action. For Rosen and, more extremely, for Robinson, it was unacceptable to overlook what they considered open anti-Semitic policies in the Soviet Union. For many, it was difficult to take a position about where the boundary should be drawn in such politically strained situations. Mercier believed scientists cannot avoid acting according to their moral values, especially in order not to repeat the mistakes of the past that allowed the rise of Nazism and all its horrors. For Bergmann, it was inadmissible to exclude eminent scholars for political reasons. Many agreed in principle with this last clear demarcation, although it should be noted that no one protested against the fact that Jordan did not receive his visa for the URSS in a similar situation. Possibly building on his experience in management positions in other international institutions, Bondi tried to define a clear boundary in agreement with the principle of free circulation accepted and implemented by the international unions. In his function of President of the ICGRG, he attempted to find a negotiating position that could preserve this principle: at least one Israeli scholar had to be invited to the Tbilisi conference, although actually participating was virtually impossible at that point.

The positions of scientists concerning these explicit political conflicts depended on many factors, including personal convictions and systems of belief, closeness to particularly affected communities, the political context in which they lived, and visions about the role and function of international scientific exchange. This last factor led to surprising outcomes if we think only in terms of the East-West dichotomy in the context of the Cold War. Those who were participating in the venture of building an international scientific community for purely scientific reasons were far more relaxed, whatever their political position, compared to those who attached greater moral values to this activity. Mercier—who tended to see the ICGRG as an ideal representation of a purely scientific community able to overcome national divides—intervened with a strong political action, boycotting the Tbilisi conference in response to the armed invasion of Czechoslovakia, that seriously undermined future collaboration. By contrast, Wheeler—who was notoriously very conservative in political matters—simply ignored the various attempts to intervene in the organization of the Tbilisi conference. He attended the conference in order to pursue his scientific interests and, a few years later, even made plans to publish a book in Russian before publishing it in English, with the aid of Ivanenko.[1]

By the same token, Wheeler was very attentive to separating the election of scholars belonging to the two different European blocs, whereas Mercier continued to insist on the creation of a unique geographical area including all of Germany as well as countries close to its linguistic and cultural sphere of influence. If we recall that Mercier was always very conscious of the need to maintain the political balance between Soviet and American scholars within the ICGRG, his insistence on creating a central European geographical area for electing members, irrespective of political divisions, suggests that Mercier deliberately wanted to make this into a political act. It is likely that under the incentive of the success of CERN, Mercier perceived the ICGRG's activities as a step toward constructing a culturally unified European scientific community. This would also explain his strong and somewhat surprising reaction to the armed attack on Czechoslovakia. His vision of an international community with a strong European component able to mediate between the two superpowers was profoundly disrupted by the Czech crisis.

---

[1]See Ivanenko to Wheeler, January 1972; Wheeler to Ivanenko, 5 April 1972; Remo Ruffini to Wheeler, 31 May 1972; Wheeler to Ivanenko, 28 July 1972; Wheeler to Ivanenko, 8 January 1973, JWP, Box 13, folder Ivanenko. The book was eventually published in English and Russian in the same year, in 1974 (Rees et al. 1974).

It is perhaps not completely coincidental that all these tensions reached a climax at about the same period. There was a mutual reinforcing of these tensions, of a cultural, social, epistemic, and political nature, which became more and more evident during the second half of the 1960s. Political disputes emerged only after the other tensions had already become explicit parts of negotiations within the ICGRG. The first major transformation occurred after the sudden formation of the field of relativistic astrophysics, which entailed a quite drastic change in the social composition of the GRG community at large as well as different views about what were the relevant research agendas to be pursued. These transformations led to a corresponding pressure to modify the structure of the ICGRG. Most ICGRG members tried to resist this pressure for a variety of reasons. These included deep epistemic disagreement as well as personal reasons related to the prestige and power some of them gained by belonging to the ICGRG. After the Tbilisi affair had made it clear that the ICGRG could not avoid explicit discussions on political matters, this led to stronger calls for change due to the evolution of the field and its changing social composition. These self-reinforcing dynamics between social, scientific, and political tensions did ultimately lead to the formation of an international democratic society as the successor body of an elite group that had pursued the unity of the "relativists throughout the World" up until that point. In the end, the very structure of the ISGRG came to embody the demand for a generational change as well as political divisions. This process led to the establishment of a rather original form of international scientific institution: a hybrid society that allowed membership of both individuals and institutions, thus allowing scientists on both sides of the Iron Curtain to participate in the decision-making body.

In this process, the issue of acceptable forms of scientific institutionalization in the Cold War context became a matter of explicit debate. Some scholars from outside the Eastern Bloc showed a surprisingly deep understanding of the political mechanisms to which their Soviet and Eastern Bloc colleagues were involved. Møller and Thorne, for instance, understood very clearly how to allow scientists from Eastern Bloc countries to participate without renouncing the principle of democratization and individual participation they wished to pursue with the transformation of the ICGRG into an international society. In their proposals, they saw that the issue of national representation might have been not as strict or dramatic as some Eastern Bloc ICGRG members stressed. In fact, it appears that in some cases scientists used this argument to settle controversies and rivalries related solely to internal power dynamics. This was the case, for instance, with the attempt by Fock and other Soviet ICGRG members to replace Ivanenko by Faddeev in 1965 and Treder's efforts to avoid Dautcourt replacing him after the GR6 conference. While disliking these dynamics, most ICGRG members from Western countries simply tried to avoid becoming involved in these "internal squabble[s]" (see Sect. 5.1).[2] Without doubt, some ICGRG members from the Soviet Union and other Eastern Bloc countries were subject to enormous constraints and political

[2]Rosenfeld to Bondi, 19 November 1968, BOND, folder 4/4A.

pressure. In other cases, however, like the instances mentioned above, the scientists involved intended to use the ICGRG to increase their status within the national socio-political hierarchy, or solve issues linked to internal dynamics of power.

Since this institutional history seems to have a somewhat singular character, one question arises: Why relativity? Was it simply the result of specific actions of particular actors having an idealistic view of international relations in the specific context of the Cold War? Or was there something in the nature of the field of GRG that inspired the actors to pioneer new modes of international cooperation?

It is not possible to answer these questions without taking into account other questions of a somewhat symmetrical character. Were the community-building activities described in this book essential to the process of the renaissance of general relativity? What was their precise role in this process? How, if at all, did the attempt to build a new community of scientists allow the emergence of what was in many respects a new field?

As I showed in Chaps. 4 and 5, the historical evolution of the institutional representation of the international GRG community can be divided into two periods. The first, spanning from the mid-1950s to the mid-1960s, could be interpreted as the formative phase, during which the initial steps were taken to institutionally unify the different research agendas under the heading of "General Relativity and Gravitation." During this period, an elite group of scientists launched a number of strategies aimed at strengthening the links between scientists working in different parts of the world and with different research agendas. This phase was mainly characterized by efforts to confront in a constructive way the epistemic uncertainty related to the loosely defined research area of GRG. The second period, from the mid-1960s to the mid-1970s, could be referred to as the maturity stage, in which scientists attached to the existing institutional structure were involved in a variety of controversies summarized above.

During the first, constructive, phase, the community-building activities played a fundamental role in the renaissance phenomenon, as interpreted by Blum, Renn, and myself (Blum et al. 2015). In the postwar period, the poorly defined field of "General Relativity, Gravitation, and related subjects" was particularly suitable for being consolidated in an international social organization because it was a mixture of various research agendas considered to be marginal compared to other major research activities (Mercier and Schaer 1962, p. 1). The number of scholars involved was simply too small to have significant relevance at the national level. When the first international conference was organized in Bern, it was explicitly an effort to build a community where there was none. The way in which this happened gives a sense of artificiality as it depended heavily on the personal decision of just one scientist, who was not even active in the field. The decision taken by an elite group of scientists to build such an institutional body unified different research agendas under this more general heading of GRG, reifying a new community of scholars. When the ICGRG was established, the formation of an elite system represented by scholars who were highly esteemed in their relevant disciplinary and national communities gave the field the air of a respectable scientific endeavor. By the same token, participation in an elite international institution helped some of the

ICGRG members strengthen the research activities in gravitation theory in their own countries. In addition, the institutionalization of GRG in the form of an elite group led to opportunities to obtain funds as well as the recognition of the GRG as a unified field in which there was an identifiable community.

This social mechanism acted at the epistemic level, too. As the field also seemed dispersed from an epistemic perspective—divided as it was in different research agendas—explicit community-building activities and a defined institutional structure were probably essential elements in bringing scientists active in different research projects into dialogue. These dynamics allowed common research projects to be identified, which in turn reinforced the diffused perception, and self-perception, of the existence of a unitary field called GRG. As we have seen, there were already tensions during the first phase, but these did not lead to any strong conflict: the general need to pursue social and epistemic unification prevailed, although some regional groups were already structuring themselves around different successful forms of collaboration.[3] However, the very fact that the ICGRG members remained attached to this international community-building project shows that they considered the international arena fundamental, despite all the possible differences and tensions.

The developments in relativistic astrophysics beginning with the discovery of quasars in 1963 completely changed the field as well as the dynamics of community building thus far stabilized and institutionalized through the ICGRG. In a sense, the incredibly swift formation of the field of relativistic astrophysics related to the First Texas Symposium had the same character of artificiality as the formation of the field of GRG arising from the Bern conference. Relativistic astrophysics emerged from a still vague intuition depending on a single discovery and the explanation of this was still very uncertain. The experts in general relativity and astrophysics who met in Texas had little in common and often had difficulty understanding each other.[4] Notwithstanding this inability to communicate effectively, the event created the field from scratch, a field that soon became enormously successful. In fact, the community that coalesced around the Texas symposia in connection with the

[3]In different countries, this process occurred at different levels, local, regional, and/or, national. In the United States, for example, an important role was played by the "Stevens Relativity Meetings" which was attended by the East Coast community of scientists working on GRG (see Appendix A.14.3). Information about these meetings can be found in Dean Rickles and Donald Salisbury, interview with Louis Witten, 17 March 2011, https://www.aip.org/history-programs/niels-bohr-library/oral-histories/36985. Accessed 12 March 2017. See also Dean Rickles and Donald Salisbury, oral interview with Jim Anderson, 19 March 2011; and Dean Rickles and Donald Salisbury, oral interview with Dieter Brill and Charles Misner, 16 March 2011. I am very grateful to Rickles and Salisbury for having given me access to the recorded interviews.

[4]One example of this is that the New Zealander mathematician Roy P. Kerr, then a postdoc at the University of Texas, presented a new set of solutions of Einstein's gravitational equations, which were later understood to be describing rotating black holes and the central mechanism for understanding quasars. When Kerr made his ten-minute presentation at the First Texas Symposium, only Papapetrou seemed to understand the importance of Kerr's findings (Thorne 1994, p. 342).

emergence of relativistic astrophysics was not identical to the community of "relativists" organized by the ICGRG. Although it was an international event, the majority of participants came from the U.S. The format of the symposia was completely different from that of the international conferences organized by the ICGRG. Furthermore, there was no attempt to build a community and no ideological attachment to the pursuit of scientific internationalism and international détente, which was extremely important in the early days of the ICGRG.

Relativistic astrophysics was so successful that eliminated two of the main needs that had led to the establishment of an international community in GRG: many scientists were soon interested in the field and the links with physics were obvious and direct, at least in terms of research agendas and goals. During the 1960s, relativistic astrophysics grew rapidly with around 600 participants present at the Third Texas Symposium. By contrast, the ICGRG tried to limit the number of its conference participants and the decision of which scientists were invited depended on factors such as the need for fair representation of both national, or local, communities and research agendas. Although it was not clear to the protagonists, the success of relativistic astrophysics could have made the existence of an international institution devoted to GRG superfluous. In 1971, Treder stated this point very clearly: general relativity was by then so entangled with many other research agendas in "normal" physics and astrophysics that the existence of an institutional representation of the GRG field had lost all the meaning it had when the field was an "esoteric" one (see Sect. 5.3). In these conditions, with strong national communities and a brand new successful field of physics, the survival of the ICGRG—or any form of international organization devoted to GRG—was far from guaranteed. The following questions then arise. Why was this organization able to survive, albeit with a profoundly modified structure? Why did scientists make so much effort to overcome the numerous difficulties when perhaps the field of research no longer needed such an international organization for its development?

Many of the reasons probably had to do with the resilience and self-perpetuation of scientific institutions. When relativistic astrophysics emerged, the ICGRG was already active and highly successful in its own right. Despite all the difficulties and differences, those involved in the decision-making within the ICGRG had the continuation of the activity as their main objective. And they almost always acted with this aim in mind. The Secretary, the Presidents and most members of the ICGRG tried to overcome the difficulties in many creative ways with the goal of preserving the institution for the simple reason that they believing it was worth saving, without referring to any real argument concerning the status of the field. Mercier, Bondi, Møller, Bergmann, Wheeler, Ivanenko, Rosen, DeWitt, and others did this for many different reasons and in different ways, but, apparently, with the same overarching target: the survival of some form of international organization. Even those who were not part of the ICGRG and wanted to modify it were actively involved in the attempt to preserve an international institution for the GRG field.

The very fact of being an "internationally recognized body,"[5] as Ivanenko often stressed, provided the ICGRG members the rationale to pursue their activities. In addition, when conflicts emerged, the structure was flexible enough to allow major modifications that maintained what members of the GRG community considered to be the spirit and the function of the institution.[6]

Linked to this motivation, which depended on the contextual factor that the ICGRG emerged and evolved before the birth of relativistic astrophysics, there are other factors related to the specific features of general relativity that allowed, first, the formation of an international community and, second, the unique outcome of the negotiations in the form of a truly international society in which participation of scientists was only partially mediated by national institutions. Together with the specific attitudes and actions of some of the main actors, this institutional process was made possible by the fact that GRG was a field with almost no direct practical application. The scientific discussions concerning the field of GRG at the international level could therefore be seen as irrelevant to military concerns.[7] This specific feature of the GRG might have been a key element that helped initiate international collaboration of a new kind, while other fields were of greater concern to political and military authorities.

In addition, the figure of Einstein represented a role model and a positive symbol for scientists working on both sides of the Iron Curtain.[8] Mercier, for instance, quoted him, along with Bohr, as an exemplary scientific figure who acted in favor of peaceful relations. Einstein's internationalist and socialist views led party officials in East Germany and other Eastern Bloc countries to explicitly refer to him as a moral compass in scientific matters. This view led in turn to the organization of many celebratory events related to general relativity and Einstein in East Germany, such as the fiftieth anniversary of general relativity in both East Berlin and Jena in 1965 (see Sect. 5.1), the centenary anniversary of Einstein's birth with a solemn festival with ceremonial addresses by the Prime Minister of the GDR and the General Director of UNESCO (Treder 1979), and the organization of the GR9 in Jena in 1980.[9] Einstein was obviously an important figure for a large section of the Jewish community working on gravitation theory at the time, and a few of the early leaders in this field—notably, Bergmann, Rosen, and Infeld—had been his close collaborators. In western culture, Einstein also had a particularly relevant place as

[5]Ivanenko to Møller, 22 July 1971, GR6P, Box A-L.

[6]Part of the argument presented here depends on the network theoretical analysis pursued by myself in collaboration with Dirk Wintergrün (Lalli and Wintergrün 2016; see also Renn et al. 2016).

[7]This is, of course, not strictly true because some of the meetings were in fact financially supported by military bodies, particularly in the United States. Scientists were able to take advantage of this situation for financing research in GRG (see Sects. 2.1 and 4.2). I maintain, however, that the possible actual applications were so tenuous that the field was not considered sensitive by military and political apparatuses.

[8]I am grateful to Dieter Hoffmann for discussions on this point.

[9]"The GRn conferences," http://www.isgrg.org/pastconfs.php. Accessed 7 March 2016.

the iconic representation of pure scientific pursuit, and more specifically of theoretical physics (Barrow 2005). All these symbolic roles Einstein held in different parts of the world, as a kind of gold standard for physicists (Hoffmann 2017) could have engendered a positive attitude toward the formation of international relations in that particular field, following Einstein's internationalist stance (Braun 2005; Rowe and Schulmann 2007).

Depending on all these factors, the construction and activities of the ICGRG and its transformation into the ISGRG was an essential step in the definition of a new field called "General Relativity and Gravitation" and the formation of an international community of "relativists." The strength of this process is exemplified by the fact that the birth of an international community of relativists happened even though some of the protagonists in the institutional venture, such as Wheeler, explicitly fought against the label of "relativists" as distinct from the rest of physics. Although Wheeler maintained these views and tried to shape the field according to this belief, in the institutional settings he acted in the best interests of the ICGRG. Particularly when political tensions arose, Wheeler helped the formation of the "relativity" community as much as Bergmann did who had been much more concerned with the establishment of this kind of community since the late 1940s.

As we have seen, there were many elements of the field of GRG and of its history that played a fundamental role in its institutional development in the international arena during the Cold War. One of these elements concerned the investiture (or self-investiture) of an otherwise marginalized professor of theoretical physics in Bern as the main pillar of the institutional body that was being built. Mercier played a major role in starting the tradition of international conferences with his idea of, and work for, the Bern conference in 1955. He assumed the role of Secretary of the ICGRG from its inception up until its transformation into the ISGRG, thus doing most of the organizational work to keep the structure going. He was involved in furthering the communication channels through editing the *Bulletin on GRG*. He was the main instigator of the transformation of the *Bulletin on GRG* into the first full-fledged scientific periodical dedicated entirely to the field of GRG despite strong opposition from authoritative members of the ICGRG. Finally, Mercier played a major role in mediating between the different positions of scholars working on opposite sides of the Iron Curtain during the most strained moments of the ICGRG. Although, as we have seen, he also certainly did take some actions that endangered the continuation of the ICGRG itself—which was then saved by negotiation by the Presidents and the good will of many participants—it is unquestionable that, from 1953 to 1977, when he finally retired from his role as Secretary of the ISGRG, Mercier had been the central actor in the path toward the GRG becoming institutionalized. By contrast, he was only a minor figure in the intellectual development of the field, as illustrated by the fact that his works in the field of GRG have almost never been cited in the scientific literature. Indeed, his contributions to the theory of general relativity were limited to stimulating philosophical interpretations that in no way influenced the major theoretical discussions revitalizing the field in the postwar period.

The case of a scientist whose main contribution to the scientific endeavor is his or her role in developing the infrastructures of science, establishing and improving communication channels, explicit community building and institutional activities is certainly not unique in the history of science. Still in the domain of 20th-century physics, other relevant examples whose active role I have explored elsewhere are Karl K. Darrow—an industrial physicist whose salary was paid by the Bell Labs to write reviews of recent advancements in physics and to serve as the Secretary of the American Physical Society (APS)—and John T. Tate, who was the editor of the APS journals from 1925 until his death in 1950, during which time he profoundly shaped the publication venues of the society by introducing important new developments (Lalli 2014, 2015, 2016). All these figures also have much in common with Henry Oldenburg—one of the first Secretaries of the Royal Society and the celebrated founding editor of *Philosophical Transactions of the Royal Society*, first published over 350 years ago and still published to this day (Hall 2002).

While the scientific production of the abovementioned scientific figures was negligible, the impact of their activities on the evolution of science was enormous. Some of the infrastructures they created, or helped develop, profoundly shaped scientific practice, and these have lasted longer than successful scientific theories and instrumental designs. Despite this substantial role, these figures have not aroused much interest among historians of science. Apart from biographical studies and more in-depth analysis focusing on Oldenburg and his network of correspondence, we still lack a historiographical framework that could help us understand the role these figures played in the evolution of modern science (Hall and Hall 1965–1986; Hall 1970; Avramov 1999).

I conclude this essay with a programmatic call. Mercier's case and its similarities with other figures that occupied an analogous niche on the interface between science and institutional administration might reveal a general pattern in how the link between scientific progress and scientific institutionalization in modern science evolves. This pattern requires closer historical scrutiny if we wish to understand the role of the institutionalization of knowledge in connection to the various phases of knowledge production, codification, and circulation in the future.

References

Avramov, Iordan. 1999. An apprenticeship in scientific communication: The early correspondence of Henry Oldenburg (1656–1663). *Notes and Records of the Royal Society of London* 53: 187–201.

Barrow, John D. 2005. Einstein as icon. *Nature* 433: 218–219. doi:10.1038/433218a.

Blum, Alexander, Roberto Lalli, and Jürgen Renn. 2015. The reinvention of general relativity: A historiographical framework for assessing one hundred years of curved space-time. *Isis* 106: 598–620.

Braun, Reiner (ed.). 2005. *Einstein-peace now!: Visions and ideas*. Weinheim: Wiley-VCH.

Hall, A. Rupert. 1970. Henry Oldenburg et les relations scientifiques au XVII siècle. *Revue d'histoire des sciences* 23: 285–304.

Hall, Marie Boas. 2002. *Henry Oldenburg: Shaping the Royal Society*. Oxford: Oxford University Press.

Hall, A. Rupert, and Marie Boas Hall (ed.). 1965–1986. *The Correspondence of Henry Oldenburg*, 13 Vols. London: Madison.

Hoffmann, Dieter. 2017. Albert Einstein (1879–1955) – relativ politisch. In *Physik, Militär und Frieden. Physiker zwischen Rüstungsforschung und Friedensbewegung*, ed. C. Forstnerm and G. Neuneck. Wiesbaden: Springer Spektrum Research. (in print).

Lalli, Roberto. 2014. A new scientific journal takes the scene: The birth of Reviews of Modern Physics. *Annalen der Physik* 526: A83–A87.

Lalli, Roberto. 2015. "The Renaissance of physics": Karl K. Darrow (1891–1982) and the dissemination of quantum theory at the Bell Telephone Laboratories. In *A bridge between conceptual frameworks: Sciences, society and technology studies*, ed. Raffaele Pisano, 249–273. Dordrecht: Springer. doi: 10.1007/978-94-017-9645-3_14.

Lalli, Roberto. 2016. "Dirty work", but someone has to do it: Howard P. Robertson and the refereeing practices of Physical Review in the 1930s. *Notes and Records: The Royal Society Journal of the History of Science* 70: 151–174.

Lalli, Roberto, and Dirk Wintergrün. 2016. Building a scientific field in the Post-WWII Era: A network analysis of the renaissance of general relativity. Invited talk at the Forschungskolloquium zur Wissenschaftsgeschichte, Technische Universität, Berlin, 15 June 2016.

Mercier, André, and Jonathan Schaer. 1962. General information. *Bulletin on General Relativity and Gravitation* 1: 1–2. doi:10.1007/BF02983127.

Rees, Martin J., Remo Ruffini, and John Archibald Wheeler. 1974. *Black holes, gravitational waves, and cosmology: An introduction to current research*. New York: Gordon and Breach.

Renn, Jürgen, Dirk Wintergrün, Roberto Lalli, Manfred Laubichler, and Matteo Valleriani. 2016. Netzwerke als Wissensspeicher. In *Die Zukunft der Wissensspeicher : Forschen, Sammeln und Vermitteln im 21 Jahrhundert*, ed. Jürgen Mittelstraßm and Ulrich Rüdiger, 35–79. München: UVK Verlagsgesellschaft Konstanz.

Rowe, David E., and Robert J. Schulmann. 2007. *Einstein on politics: His private thoughts and public stands on nationalism, Zionism, war, peace, and the bomb*. Princeton: Princeton University Press.

Thorne, Kip S. 1994. *Black holes and time warps: Einstein's outrageous legacy*. New York: WWNorton.

Treder, Jürgen (ed.). 1979. *Einstein-Centenarium 1979*. Berlin: Akademie-Verlag.

Appendix A
Research Centers on Fields Related to General Relativity Around the Mid-1950s

In this Appendix, I will give short descriptions of the various research centers active in the mid-1950s that were working on topics soon to be included in the larger GRG domain. As discussed in Chap. 2, I use a broad definition of research centers. These are institutions (such as universities, research institutes, sections of national academies of science, etc.) in which there was at least one principal investigator who had an institutional position stable enough to attract postdocs and/or produce new Ph.D.'s in the field. In the presentation that follows, research centers are organized according to their national contexts. This is intended to convey the status of the relevant research activities in the various countries at the outset of the renaissance process, which coincided with the community-building activities in the international arena discussed in this book. The list presented is broad, but it does not claim to be exhaustive. It should not be considered to be a complete representation of what was happening in the different countries. Research activities that did not have a relevant part in the formation of the community might have escaped the author's attention and research centers that were established after 1956 have been excluded. Priority has, in fact, been given to research activities that were represented in the ICGRG when it was first established in 1959. It is assumed that these research centers were considered by the emerging community to be the most active or relevant for topics related to general relativity in the late 1950s.

It was not easy to choose the order of presentation of research centers. The argument of this work and considerations about the different relevance of the centers would have suggested beginning with the status of these centers in the United States and the Soviet Union. However, the following system has been used to make it easier to search within the document: countries are presented in alphabetical order, while research centers within each country are roughly in chronological order based on the date they were established. The scientists' dates of birth and death have only been included for scientists who were in charge of these research centers in the mid-1950s.

R. Lalli, *Building the General Relativity and Gravitation Community During the Cold War*, SpringerBriefs in History of Science and Technology,
DOI 10.1007/978-3-319-54654-4

A.1 Belgium

Free University of Brussels

Within a tradition firmly linked to the French school in differential geometry, at the time of the Bern conference, two Belgian scientists were working on mathematical aspects of the theory of general relativity at the Free University of Brussels: mathematician Robert Debever (1915–1998) and mathematical physicist Jules Géhéniau (1909–1991). Around the mid-1950s, Debever and Géhéniau initiated a fruitful collaboration on a research agenda concerning the invariants of the Riemann tensor. Notwithstanding these favorable preconditions, the center in Brussels did not develop further in the period between the mid-1950s and the mid-1960s. It was essentially limited to the work of Debever and Géhéniau, who were also both pursuing active research in other topics not directly related to GRG. During this decade, the only Ph.D. student working in this field was Michel Cahen. He earned his Ph.D. in 1960 with a dissertation on the unified theory of gravitational and electromagnetic fields related to the equations of Rainich and Wheeler. The capacity to attract postdocs from abroad was also limited, perhaps also as a consequence of the refusal by the U.S. Air Force to financially and logistically support this center for political reasons (Goldberg 1992).[1] However, the center organized one of the first smaller European international meetings in the late 1950s, which was attended by researchers of the younger generation.[2] Neither Debever nor Géhéniau became a member of the ICGRG.

A.2 Denmark

Institute for Theoretical Physics, Copenhagen University

Authoritative theoretical physicist Christian Møller (1904–1980) was a central figure in the development of theoretical physics in Denmark and played a leading role in the Danish involvement in establishing CERN in the mid-1950s (Blum and Hartz 2017). At the same time, Møller was one of the few theoretical physicists of

[1]The reason was that Géhéniau was identified with the Belgian Communist party. Joshua Goldberg, e-mail to the author, 3 March 2016; and Goldberg, Joshua, 21 March 2011, interview with Donald Salisbury and Dean Rickles. https://www.aip.org/history-programs/niels-bohr-library/oral-histories/34461. Accessed 5 March 2016.

[2]Goldberg, Joshua, 21 March 2011, interview with Donald Salisbury and Dean Rickles. https://www.aip.org/history-programs/niels-bohr-library/oral-histories/34461. Accessed 5 March 2016.

the older generation who was actively engaged in research on gravitation theory in the 1950s. In 1952, he had published a textbook on general relativity, which at the time of the Bern conference was considered to be one of the most up to date (Møller 1952).[3] As a research center, the Institute for Theoretical Physics was characterized by its role as a meeting point for scholars coming from abroad for both short and long periods, rather than by training new Ph.D. students in the field. Thus, it assumed a particular relevant role in improving links between the various groups working on different topics related to GRG. Among the researchers who spent a period at the Institute for Theoretical Physics in the period 1955–1965 are Stanley Deser, Bryce S. DeWitt, Leopold E. Halpern, Bernard Jouvet, Oskar Klein, Arthur B. Komar, Bertel Laurent, Charles W. Misner, Erwin Schrödinger, and Frank R. Tangherlini.

One important event taking place in this institutional setting was the month-long meeting dedicated to the quantization of the gravitational field, which was held in July 1957, only a few months after the Chapel Hill conference (Blum and Hartz 2017, DeWitt 2017). In 1957, Møller also became the Director of the newly established Nordic Institute for Theoretical Physics (NORDITA) and, one year later, Léon Rosenfeld was also invited to join the institute. The Copenhagen research environment had a strong representation within the ICGRG from the outset: both Møller and Rosenfeld were among the founding members of the ICGRG in 1959.

A.3 Federal Republic of Germany (West Germany)[4]

A.3.1 Hamburg University and Hamburg Observatory

In the mid-1950s, Pascual Jordan (1902–1980) was without doubt the most authoritative scientist working on research connected to general relativity in West Germany. A former active member of the Nazi party, he was not able to obtain a guest professorship at the University of Hamburg until 1947 after the denazification process ended and thanks to the personal recommendation of Wolfgang Pauli (Beyler 1996; Hoffmann and Walker 2007). At this university, he worked in fields related to general relativity with a focus on two areas: Dirac's large numbers hypothesis and the formulation of a scalar-tensor theory of gravitation (Goenner 2012). Jordan's alternative gravitation and cosmological theories had a certain

---

[3]Wheeler, for instance, refers in his notebooks to Møller's book as the major source of reference in his attempt to learn the theory. Relativity Notebook 1, JWP, Box 39 (quoted in Blum 2016).

[4]For a detailed description of the development of the field of GRG in Germany, see Goenner (2016).

impact on other researchers in West Germany outside his group. After he became a full professor in Hamburg, he was able to start building a stronger research group with his students and later with postdocs and more senior researchers coming from abroad. By 1965, those who had obtained a Ph.D. under Jordan's supervision included some of most influential German experts of general relativity over the following decades: Engelbert Schucking (Ph.D. in 1956), Jürgen Ehlers (Ph.D. in 1958), Wolfgang Kundt (Ph.D. in 1958), and Manfred Truemper (Ph.D. in 1962). Jordan collaborated very closely with all his students and the focus of the group rapidly shifted toward topics more directly associated with general relativity proper. The work Jordan did with Kundt and Ehlers on the exact solutions of Einstein's field equation, for instance, had a particularly strong impact on the renaissance of general relativity. Among those coming from abroad who spent a period of research in Hamburg in the same timeframe, the following are worth mentioning: Peter Bergmann, Dieter Brill, Joshua Goldberg, Wolfgang Rindler, Ivor Robinson, and Rainer Sachs. This center was one of the few non-American centers to be financially supported by the U.S. Air Force program in gravitation.

Closely associated to Jordan's emerging group in Hamburg, the Director of the Hamburg Observatory, Otto Heckmann (1901–1983), was also working on research questions related to general relativity, particularly in the field of cosmology. Heckmann and Jordan actively collaborated and shared co-workers. Schucking, in particular, even before earning his Ph.D. with Jordan, became Heckmann's assistant and remained his closest collaborator up until 1959 when he moved to the United States. Ehlers, too, cooperated closely with Heckmann after having earned his Ph. D. Given the strong links between these two groups, we could consider the Hamburg environment to be a large research center with two leading figures, with Jordan's group certainly the largest. Heckmann, however, also had at least one Ph. D. student in the field: the Hungarian Istvan Ozsvath (Ph.D. in 1960).

A.3.2 Institute for Theoretical Physics, Freiburg University

In Freiburg, a much smaller group was being established around the figure of Helmut Hönl (1903–1981), an eminent theoretical physicist who had contributed in particular to the development of quantum mechanics. He first became interested in the theory of general relativity in the late 1930s, but by the time of the Bern conference he had only published a few papers on the subject in collaboration with August W. Maue. The group started to grow as of the second half of the 1950s, when Hönl supervised a few Ph.D. students on topics within the field of GRG, including Charlotte Soergel-Fabricius (Ph.D. in 1960), Heinz Dehnen (Ph.D. in 1961), and Hubert Goenner (Ph.D. in 1966). In the early 1960s, Konrad Westpfahl, another senior member of the institute, began collaborating on this research.

A.4 France

Collége de France and Institut Henri Poincaré, Centre National de la Recherche Scientifique (CNRS), Paris

In the mid-1950s, two research centers had been already established in Paris. These two groups were based at the two different universities (the Collége de France and the Sorbonne), but they also both had an institutional link to the CNRS. Mathematician Andre Lichnerowicz (1915–1988) had been working on the Cauchy problem in general relativity since his doctoral studies under Georges Darmois in the late 1930s, where he made considerable progress on the initial value problem in general relativity following lines of research initially pursued by Élie Cartan and Darmois himself (Lichnerowicz 1992; Stachel 1992; Choquet-Bruhat 2014). From 1949 to 1952, he was Professor of Mathematical Methods at the University of Paris and, in 1952, he obtained a chair at the Collége de France, where he introduced a course on general relativity. From 1949,, he trained many students working on different aspects of the theory of general relativity, such as mathematical problems of Einstein's theory, alternative theories of gravitation and the unified theory of gravitation and electromagnetism. By 1965, Lichnerowicz had been the supervisor of an impressive number of dissertations on these topics, including those of Yves Thiry in 1950, Yvonne Bruhat in 1951,[5] Pham Mau Quan in 1954, Josette Charles in 1956, Francoise Maurer in 1957, Francoise Hennequin in 1958, Pierre V. Grosjean in 1958, Louis Bel in 1960, Cahen (as co-supervisor together with Debever) in 1960 (see Appendix A.1), Albert Crumeyrolle in 1961, Robert Vallee in 1961, Marcel Lenoir in 1962, and Jean Vaillant in 1964 (Goenner 2014).

The second group active in Paris was led by French theoretical physicist Marie-Antoinette Tonnelat (1912–1980). Tonnelat had earned a doctorate under the supervision of Nobel laureate Louis de Broglie in 1939 with a dissertation on the theory of photons in a Riemannian space. After having worked in the group directed by de Broglie, she was able to start her own group in 1945, when she became Director of Research at the CNRS in Paris. While she had already initiated her program on unified field theory, in establishing the research topic of her group, she was probably influenced by the approach followed by Erwin Schrödinger toward a pure-affine theory, with which she became acquainted when spending a year at DIAS (see Appendix A.7). Like Lichnerowicz, Tonnelat also had a good track record of dissertation supervision within the field of GRG. By 1965, the following

[5]Yvonne Bruhat was known by the surname Fourès-Bruhat while she was married to mathematician Leonce Fourès but later changed her surname to Choquet-Bruhat after her second marriage (to mathematician Gustave Choquet in 1961).

had all earned Ph.D.'s under her guidance or with her support: Jacques Levy in 1957, Pam Tan Hoang in 1957, S. Kichenassamy in 1958, Jean Hely in 1959, Marcel Bray in 1960, Liane Bouche in 1961, Nguyen Phong-Chau in 1963, Philippe Droz-Vincent in 1963, Aline Surin in 1963, Sylvie Lederer in 1964. Under Tonnelat's influence and in collaboration with her, Stamatia Mavrides and Judith Winogradzki also worked on the program of unified field theory in the late 1950s (Goenner 2014).

From the perspective of community building, these two groups played a major role in training a new generation of experts, particularly of French nationality, and in sparking research on similar topics in other European countries, such as Italy (see Appendix A.9). However, it appears that they did not act as strong centers for international postdoctoral studies. In addition, the interconnections between the two groups were quite weak. Collaboration seems to have been mainly of a formal and organizational nature, rather than based on joint research projects. The major collaborative venture was organizing the Royaumont conference, which played a key role in the establishment of the ICGRG and in strengthening contacts between Eastern and Western scholars (see Sects. 4.2 and 4.4). Like Copenhagen, Paris was also strongly represented within the ICGRG. Both Lichnerowicz and Tonnelat became members of the ICGRG when it was established during the Royaumont conference and both assumed the role of co-President for the subsequent three years.

A.5 German Democratic Republic (East Germany)

Institute for Pure Mathematics, German Academy of Sciences at Berlin

In East Germany, there was one major active group working on general relativity and related topics in the mid-1950s. This group was formed around the figure of the Greek theoretical physicist Achilles Papapetrou (1907–1997) at the Institute for Pure Mathematics of the German Academy of Science at Berlin (DAWB). Papapetrou had been working on a variety of topics related to general relativity since the late 1930s. Known for his left-wing attitudes, he had to leave Greece at the dawn of the Greek civil war. He first went to Dublin, where he worked with Schrödinger between 1946 and 1948, on the unified field theory program he pursued at the Dublin Institute for Advanced Studies (DIAS) (Goenner 2014). He later spent almost four years with Rosenfeld at the University of Manchester (see Appendix A.13.4) where he conducted research notably on the equations of motion in general relativity. In 1952, Papapetrou went to East Germany where he established a research group on general relativity at the DAWB (Hoffmann 2017). The main rationale for establishing this group was that it was intended to be linked directly to the solar eclipse expeditions planned by the German Academy of Sciences under the direction of Erwin Finlay-Freundlich, an early associate of Albert Einstein's and observer at the astrophysical observatory in Potsdam (the Einstein Tower) from its establishment in 1920 up until 1933, when he was removed from office after the implementation of the anti-Semitic Civil Service Law by the Nazi regime (Hentschel 1997). In addition, it was hoped that Papapetrou would re-establish the Berlin research tradition in general relativity, which had been interrupted during the Nazi regime. In the period when Papapetrou worked in Berlin, from 1952 to 1961, he trained at least two Ph.D. students on topics in the field of GRG: Hans-Jürgen Treder (Ph.D. in 1956) and Georg Dautcourt (Ph.D. in 1962). When Papapetrou left the GDR in 1961 to take up the position of Director of Research at the CNRS in Paris, he was replaced at the DAWB by his former student, Treder. In the future, Treder would have a strong influence in shaping and strengthening the field in the GDR also because of his leading role in science administration, which was not only rooted in his scientific competence, but was also backed by the political authorities. None of the scientists working in the GDR became members of the ICGRG when it was established. The first scientist to become a member of the ICGRG was Treder in 1969 (see Sect. 5.1).

A.6 India

Despite the absence of Indian scientists within the membership of the ICGRG up until 1969, India hosted a fairly considerable number of research activities in the mid-1590s, at least by standards at the time. For a variety of reasons that I will not address here, these activities were pursued in isolation from the increasing network of scientists who were establishing a community in North America and Europe (Lalli and Wintergrün 2016). The most relevant exponent in this tradition was mathematical physicist Vishnu Vasudev Narlikar (1908–1991). After having completed his studies at Cambridge University, he became Professor of Mathematics at **Banaras Hindu University** in Varanasi in 1932 and pioneered research on general relativity in India on subjects such as exact solutions of general relativity, unified field theory of electromagnetism and gravitation, equations of motion in general relativity, and invariants of Riemannian metrics. One of his Ph.D. students was Prahalad Chunnilal Vaidya (Ph.D. in 1947), who by the mid-1950s had already established a second flourishing research center on GRG at **Gujarat University**. Narlikar trained, and collaborated with, many younger Indian researchers and had a long-term influence on them. Probably inspired by his research, his son, Jayant Vishnu Narlikar would become an important astrophysicist and cosmologist. Vaidya would be the first Indian scientist to join the ICGRG, which he did in 1969, when the Committee on Gravitation was established in India (see Sect. 5.1).

A.7 Ireland

Dublin Institute for Advanced Studies (DIAS)

By the mid-1950s, DIAS was certainly one the largest and most authoritative research centers in fields related to general relativity and gravitation in Europe. Three senior scientists were pursuing different kinds of studies in this area. The most influential in community building and in establishing international collaboration was perhaps the Irish mathematician John L. Synge. Synge had been working on general relativity throughout his entire career since the early 1920s. During the low-water-mark period, he established himself as one of the greatest authorities on the mathematical problems of general relativity. After almost twenty years at the University of Toronto, Synge agreed to come back to Ireland as Senior Professor at DIAS. There, he dedicated himself almost completely to research in relativity theories, producing two important books, one on the special theory of relativity (Synge 1956) and the other on the general theory (Synge 1960) and collaborating with a number of younger researchers, including Anadijiban Das, Petros Serghiou Florides, Lochlainn O'Raifeartaigh, and Felix Pirani. He also supervised the dissertation of Werner Israel (in 1960) who would later contribute significantly to the theory of black holes.

The other two scientists who were active at DIAS at the time were Nobel laureate Erwin Schrödinger (1887–1961) who had been pursuing a research program on the unified field theory of gravitation and electromagnetism since the early 1940s, and the Hungarian mathematical physicist Cornelius Lanczos (1893–1974), who was one of the greatest authorities on general relativity during the low-water-mark phase. Before he returned to Vienna in 1956, Schrödinger had the opportunity to work with many colleagues at DIAS, including Bertotti, Papapetrou, and Tonnelat, who would all play a significant role in the renaissance process and in the activities related to establishing the international community. It seems that Lanczos did not play as great a role in establishing collaboration with younger researchers active in the renaissance process. Of the three, Synge was the scientific figure who exerted a long-term influence on the research activities in the field both at the local and international level. This was institutionally recognized during the foundation of the ICGRG when Synge was invited to become a member.

A.8 Israel

Israel Institute of Technology—Technion, Haifa

The American-Israeli theoretical physicist Nathan Rosen (1909–1995) had worked as Albert Einstein's assistant at the Institute for Advanced Studies (IAS) in Princeton between 1934 and 1936. With Einstein, Rosen produced a number of achievements, including the elaboration of the Einstein-Rosen space-time bridge published in 1935, the formulation of the EPR paradox (the Einstein-Podolsky-Rosen criticism of quantum mechanics) in the same year, and the paper on gravitational radiation published in a modified version by Einstein alone after the end of Rosen's stay at IAS (Kennefick 2007). After working for two years at the University of Kiev in the Soviet Union, Rosen returned to the United States, but eventually he moved to Israel in 1953 to take up a professorship at the Technion in Haifa. Besides playing an important role in the growth of the institute by serving in many administrative positions, Rosen established a small research center on the theory of gravitation there. Unlike many other centers, at that time, Rosen's center was pursuing research in general relativity proper, focusing in particular on gravitational waves and the equations of motion. At Technion, between 1953 and 1965, Rosen trained a new generation of Israeli scholars pursuing research in the field of GRG, including Asher Peres (Ph.D. in 1960), Gidon Erez (Ph.D. in 1960), and Moshe Carmeli (Ph.D. in 1964). Moreover, he had a considerable influence on the British mathematical physicist Ivor Robinson. The center would grow further after 1959 when American theoretical physicist and general relativity expert Gerald E. Tauber joined the Technion. Rosen was certainly considered one of the greatest authorities in general relativity in the 1950s and became a member of the ICGRG when it was founded.

A.9 Italy

Notwithstanding the very strong mathematical tradition in the field of tensor calculus and differential geometry that had a profound impact on the origin and development of general relativity, by the mid-1950s, research on general relativity in Italy was at a standstill. This form of research had been almost completely disrupted by political events during the low-water-mark period, first with the fascist academic reforms in the early 1930s and, most dramatically, by the Racial Laws heralded in 1938. The Italian mathematical community suffered from an intensified political influence of the fascist regime on academic and scientific life. One of the most active mathematicians in the field during the 1930s was still Tullio Levi-Civita, whose work had a tremendous impact on the genesis of general relativity along with that of his teacher Gregorio Ricci-Curbastro. In 1938, the Racial Laws abruptly ended his career and his chances of pursuing active research. The impact of this political disruption of scientific activity, which also led to the exile of Enrico Fermi, unanimously considered the father of theoretical physics in Italy, and other physicists was still evident in the mid-1950s (see, e.g., Goodstein 1982; Nastasi and Tazzioli 2005; Bergia 2005).

In the fields connected to general relativity, the only institutional entity that might be considered as a research center active in that period was the group under the leadership of Bruno Finzi (1899–1974) at the **Polytechnic Institute of Milan**.

Finzi was a Jewish mathematician and engineer who had been strongly influenced by the work of Levi-Civita. Since the late 1940s, he had been working in areas related to general relativity in collaboration with other scholars, both senior and younger ones, including Emilio Clauser, Paolo Udeschini and, in particular, Maria Pastori. While the group was quite sizeable, this research center remained isolated all through the renaissance, possibly also because they continued to publish in Italian and were not involved in community-building activities (Goenner 2014).

Another group that was established in Rome as late as 1957 had a much greater relevance for building the larger international institutional framework. Mathematician Carlo Cattaneo (1911–1979)—a former student of Levi-Civita's—had been working on classical mechanics and fluid mechanics before he turned his interest toward mathematical methods in the theory of general relativity in 1957, when international community-building activities were already under way. After he became a professor at the **University of Rome** in 1957, he was able to establish a group on general relativity that was instrumental in the growth of interest in the theory of general relativity in Italy. This happened in 1960, when a research group called "Einstein's theory of gravitation" was established within the Italian National Research Council, which, besides Cattaneo, included Silvano Bonazzola, Mario Castagnino, and Giorgio Ferrarese. In the 1960s, along with the astrophysics group established by Livio Gratton at the Laboratorio Gas Ionizzati in 1960, Cattaneo and his research group contributed to initiating research on relativistic astrophysics in Italy. One of the first results of this was the work of Franco Pacini together with Bonazzola, which would have a strong impact on the study of neutron stars. Cattaneo's decision to pursue research in general relativity was motivated by his close intellectual and personal relationship with Lichnerowicz. This connection probably also facilitated Cattaneo's rapid entry into the international community. Cattaneo, indeed, immediately became a member of the ICGRG when it was first established (see Bonolis et al. 2017).

A.10 Poland

Institute of Theoretical Physics, Warsaw University

Like Rosen, Polish Jewish theoretical physicist Leopold Infeld had also acquired a reputation as a former collaborator of Albert Einstein in the 1930s. Infeld had replaced Rosen as Einstein's assistant at IAS in 1936 and remained in this position until 1939. In this period, Infeld's biggest achievement was probably formulating, together with Einstein and Banesh Hoffmann, the Einstein-Infeld-Hoffmann equations of motion in general relativity in 1938, which is considered one of the greatest advances in general relativity during the low-water-mark phase (Einstein et al. 1938).

After his period of cooperation with Einstein and thanks to the support of the authoritative American mathematical physicist Howard P. Robertson, Infeld obtained a professorship at the University of Toronto in 1940, where Synge was a professor (Lalli 2016). After World War II, the decline in political relations between the United States and the Soviet Union with the formation of two geopolitical blocs had a dramatic impact on Infeld's career. Although it was not as strong as the Red Scare in the United States, Canada also experienced growing hysteria against the danger posed by spies and leftists present in the country. Because of his political leaning toward peaceful cooperation and demilitarization as well as his nationality, rumors spread that Infeld was serving as a Soviet spy by passing nuclear secrets to Eastern Bloc countries. Demoralized by this unfair campaign against him and following his desire to work on rebuilding Polish science after the damage during World War II, Infeld decided to leave Canada and moved permanently to Poland in 1950. He soon became a member of the Polish Academy of Sciences, which had just been established, as well as Head of the Institute of Theoretical Physics at the University of Warsaw. From 1950 up until his death in 1968, Infeld made major contributions to the growth of Polish physics and, in return, he enjoyed enormous freedom to pursue and promote his own research interests. Under his direction, the Warsaw center then grew into one of the most prominent in the world and, around the mid-1950s, was certainly the strongest center in countries under Soviet influence.

Infeld began long-term collaboration with theoretical physicist Jerzy Plebanski and trained a number of Ph.D. students, including Andrzej Trautman (Ph.D. in 1959), Stanislaw Bazanski (Ph.D. in 1959), Roza Michalska (Ph.D. in 1966). One of Infeld's explicit goals was to increase the international prestige of the Polish Academy of Sciences and, more specifically, of his center. For this reason, he pushed his students and colleagues to publish in the scientific journal of the Academy and, at the same time, established a policy of openness by inviting international guests to pursue research in Warsaw.[6] By 1965, the University of Warsaw had hosted a considerable number of emerging experts in the field of GRG, including Pirani, Robinson, and John Stachel. It was the first and, up until the 1960s, only research center in the Eastern Bloc to become part of the increasing network of centers for postdoctoral pilgrimage, which was one of the biggest differences of the renaissance process compared to the previous period. Together with his students and collaborators, Infeld also organized the first international conference on GRG held in the Eastern Bloc: the 1962 GR3 conference in Warsaw and Jablonna (see Sect. 4.4). At the time of the Bern conference, Infeld was certainly considered one of the greatest experts in the physical aspects of the theory of general relativity and was among the founding members of the ICGRG in 1959. At that time, he was the only non-Soviet representative of Eastern Bloc countries.

[6]Andrzej Trautman, interview with Donald Salisbury, 27 June 2016, to appear in *EPJH*. I am grateful to Salisbury for making the content of this interview available.

A.11 Sweden

Stockholm University

Since his 1926 ground-breaking contribution to the five-dimensional unified theory of gravitation and electromagnetism first formulated by Theodor Kaluza, Oskar Klein (1894–1977) had become one of the most prominent Swedish mathematical physicists. After more than a decade abroad, Klein accepted a chair at Stockholm University in 1930. While contributing significantly to various theoretical developments in quantum mechanics and nuclear physics, his interest in approaches linking quantum physics and general relativity continued to be significant. As far as original research is concerned, he was not very active in the field of GRG in the mid-1950s, and the research center was rather small. Between the mid-1950s and the 1960s, only one Ph.D. student completed a dissertation in GRG: Bertel Laurent in 1959. The center also only attracted a few scholars from abroad; notably, Ivor Robinson and Leopold Halpern. Nonetheless, at the outset of the renaissance process, Klein was perceived as one the major authorities and the Stockholm center was considered to be one of very few at the time, particularly for research into quantization of gravitation. Stockholm was in fact quoted by DeWitt in his list of the eight centers working on the theory of gravitation at the end of 1955 (DeWitt 1957). Klein was one of the founding members of the ICGRG.

A.12 Switzerland

University of Bern

Given the relevant role Switzerland would play in bringing about international collaboration in the emerging field of GRG and in establishing and developing the ICGRG, one might perhaps expect to find a quite active group based at the University of Bern. Yet this was not the case. As discussed at length earlier, André Mercier was mostly interested in the philosophical implications of the theory of general relativity, and was not pursuing any research agenda in this field in the mid-1950s, and neither was Pauli (see Sect. 4.1). A clear indication of this lack of activity is that no presentations by Swiss scholars were given at the Bern conference in 1955. It should be added, however, that once the ICGRG was established, Mercier's international connections and his role in the ICGRG helped him strengthen the research activities in theoretical physics, and particularly in the field of GRG, in Switzerland, especially in Bern. In any case, the University of Bern should not be seen as a research center in this early phase of the renaissance. Mercier was the main driving force in the organization of the Bern conference in 1955, a founding member of the ICGRG and its Secretary from when it was established until after the ISGRG was created to replace the ICGRG.

A.13 United Kingdom

A.13.1 Cambridge University

At Cambridge University, research activities connected to general relativity began back in the late 1940s. In 1948, the astronomers Fred Hoyle and Thomas Gold, and the mathematician Hermann Bondi, all with teaching positions at Cambridge, formulated the Steady State theory of cosmology, a theory of the universe that would compete with the expending universe model up until the late 1960s (Kragh 1996). Several students were attracted by these new developments concerning the links between the theory of gravitation and cosmology on which Bondi, Gold, and Hoyle were working. In the same period, the famous Nobel laureate, theoretical physicist Paul A. M. Dirac began developing the Hamiltonian formulism, which was considered by others to be an important step in the path toward the quantization of Einstein's equation. In the late 1930s, Dirac had also proposed an intriguing idea concerning relations between the age of the universe and the gravitational constant and the mass of the universe, respectively. While broadly discredited, the large numbers hypothesis, as Dirac's bold cosmological idea was called, was still generating a certain amount of interest. While Dirac would not personally start working on topics related to general relativity until the late 1950s, he was well disposed about these kinds of topics. Dirac's support and the fascinating personalities of the younger Bondi, Gold and Hoyle created an intellectual climate that attracted students to reflect on topics related to gravitation and cosmology.

Despite the large number of people interested in GRG, it is not possible to say that at Cambridge in the mid-1950s there was a center working on specific research agendas connected to general relativity, as was the case in the United States and other countries. The abovementioned scientists were working on a variety of different topics. Only Dirac and Bondi proposed at least one dissertation project each connected to general relativity. Dirac's Ph.D. student was Dennis Sciama (Ph.D. in 1953), while Pirani was working with Bondi on a second Ph.D. (earned in 1956), both concerning Mach's principle. The intellectual climate, however, led a few students to enter the field during the 1950s, notably, Roger Penrose (Ph.D. in 1957) inspired by Sciama, and Roy Kerr (Ph.D. in 1959). After Sciama became a professor at Cambridge in 1961, the tradition of research was stabilized through the creation of a well-recognized, important research center explicitly devoted to general relativity and cosmology, which would train a new generation of relativists, including George Ellis (in 1964) Stephen Hawking (in 1966), and Brandon Carter (in 1967). Dirac was among the founding members of the ICGRG when it was established.

A.13.2 King's College London

In 1954, Hermann Bondi was offered a professorship in applied mathematics at King's College London. Only then did Bondi set up the first research group working with a clear focus on gravitational theory, together with his Ph.D. student Pirani and British mathematician Clive Kilmister, who was already working on the mathematical aspects of physical theories. The group was called King's Gravitational Theory Group. After the Bern conference, Bondi's major research focus became gravitational radiation theory, which he pursued with a number of students and postdocs, making King's College one of the most relevant research centers in the world for this type of research. Between the mid-1950s and the mid-1960s, several young scientists pursued research on this topic at King's College, often in close cooperation with Bondi and his group, including Leslie Marder, Trautman, Rainer Sachs, Goldberg, Penrose, Ted Newman, Wolfgang Rindler, Alfred Metzner, Sciama, George Szekeres, Robinson, and Chris Collinson. The group also began organizing frequent, successful meetings and workshops dedicated to the theory of general relativity, attended by all the community of scientists working in the field in London. Bondi was one of the founding members of the ICGRG after he proposed the idea of establishing the committee.

A.13.3 Other Colleges at the University of London

In other colleges of the University of London, various scientists were pursuing more isolated activities in the field of cosmology and general relativity in the mid-1950s. The most relevant was probably the **Imperial College**, where Gerald J. Whitrow had been a lecturer in applied mathematics since 1945. He was pursuing research in relativity and cosmology and trained Ph.D. students such as Charles Rayner (Ph.D. in 1954) and Rindler (Ph.D. in 1956). By the mid-1960s, the Imperial College had also hosted a number of visiting specialists for shorter periods, including Hans Buchdahl, Peter Higgs, Geoffrey Stephenson, and Trautman.

At the **Royal Holloway College**, the authoritative astronomer and cosmologist William H. McCrea was continuing to pursue his research on cosmological problems and trained Ph.D. students in the field, including Jack Hogarth (Ph.D. in 1953), William Davidson (Ph.D. in 1959), and Petros Florides (Ph.D. in 1960). In terms of size and amount of activity, these other groups did not have the same strength or the same impact as the King's College research group. This was also reflected in the composition of the ICGRG. Bondi was one of the founding members in 1959 and Kilmister was the third British expert to become member when the ICGRG was enlarged in 1965. However, the parallel work of different groups in London meant more opportunities for interaction, which bore fruit when the King's College group started organizing larger activities around a common area of interest.

A.13.4 Other British research centers

There were other places in the United Kingdom where work in the field was pursued in the mid-1950s. It is worth mentioning in particular **Manchester University**, where authoritative physicist Léon Rosenfeld was a professor from 1947 to 1958 and the **University of Leeds**, home institution of the British mathematician Harold S. Ruse, who had been interested in the mathematical developments of general relativity since the early 1930s. None of these other groups had the same stability as research centers in the field of GRG as Cambridge and King's College had between the 1950s and the early 1960s. However, in the early 1950s, the University of Manchester was perceived as one of the active research centers conducting research on quantization of the gravitational field and the equation of motion in general relativity. Rosenfeld had been a precursor of this research area with the first paper in the field in 1930 (Blum and Rickles 2017, Rosenfeld 2017, Salisbury and Sundermeyer 2017). In the early 1950s, his involvement with these types of question seemed to be weak, but, between the late 1940s and the mid-1950s, he supported the work in this area of the young Indian-born theoretical physicist Suraj N. Gupta and that of his former Ph.D. student Ernesto Corinaldesi in addition to research on the equation of motion in general relativity conducted by the Greek theoretical physicist Achilles Papapetrou (Blum and Hartz 2017). For these reasons and for his expertise in the quantization problems, Rosenfeld was seen as a natural candidate to join ICGRG when it was established, even though he was not particularly active at the time. In any case, Rosenfeld was already a member of NORDITA in Copenhagen when the ICGRG was established.

A.14 United States

A.14.1 Syracuse University

Peter Bergmann (1915–2002) was a close collaborator of Albert Einstein's from his arrival in the United States as a German Jewish refugee in 1936 up until 1941. In 1947, he was appointed Assistant Professor of Physics at the University of Syracuse and soon inaugurated one of the first physics research centers specifically dedicated to the theory of gravitation in the world. The main research program of the Syracuse center was directed toward the unification of general relativity with quantum mechanics (Salisbury 2012). The approach developed by Bergmann in collaboration with his students and co-workers was the non-perturbative canonical quantization of Einstein's equation. From the late 1940s to the early 1960s, Bergmann was the supervisor of many Ph.D.'s on subjects connected to general relativity and gravitation, including those of Henry Zatkis in 1950, Ralph Schiller in 1952, Robert Penfield in 1952, James L. (Jim) Anderson in 1952, Joshua Goldberg in 1952, Ezra

Ted Newman in 1956, Allen Janis in 1957, Irwin Goldberg in 1957, Rainer Sachs in 1958, and John Boardman in 1962. Many of them continued to be active in the field of general relativity and gravitation. Collaboration between these younger scientists and Bergmann often continued after graduation, and some of them managed to create new research groups.

From the second half of the 1950s, Syracuse rapidly grew as one of the biggest centers in GRG, thanks to new appointments and by attracting a number of visitors as postdocs or visiting scholars who stayed at the university for short or long periods. Those who worked at Syracuse on subjects linked to GRG in the 1950s and the early 1960s include Arthur Komar from 1957 to 1965, Roy P. Kerr from 1959 to 1960, Wolfgang Kundt in 1959, Ivor Robinson from 1960 to 1962, Roger Penrose from 1960 to 1961, Engelbert Schucking in 1961, Andrzej Trautman in 1959 to 1960 and again in 1961, Jürgen Ehlers from 1962 to 1964.[7] Richard Arnowitt also stayed at Syracuse University as an assistant professor and later associate professor from 1956 to 1959, when he began working with Stanley Deser and Charles Misner on the Hamiltonian formulation of general relativity, which would become known as the ADM formalism. This brief summary shows that Syracuse was one of the major centers in the renaissance process, and not only for the work done by its founder, Peter Bergmann, and his group, but also because it served as a node in the growing network of scientists working on various topics within the field of GRG. Bergmann was among the founding members of the ICGRG when it was established in 1959.

A.14.2 Princeton University

During the renaissance process, Princeton University was comparable with Syracuse in terms of importance as a research center on the relativistic theory of gravitation. The starting point for establishing this research center was the decision by the authoritative nuclear physicist John A. Wheeler (1911–2008) to begin teaching, and at the same time learning about general relativity in 1952. Wheeler initiated this program immediately after his strong involvement in national defense research and in the development of the hydrogen bomb ended with the preparation of the first test of the thermonuclear device Ivy Mike. The group forming around Wheeler at Princeton became one of the most active groups in the field from the mid-1950s onward (Wheeler and Ford 1998). Initially, his major area of research was directed at developing a theoretical model able to describe the behavior of particles in term of fields, through the concept of gravitational electromagnetic entities, called geons (Blum 2016). Among those who obtained a Ph.D. under

[7]For more detailed information about this research center, see Goldberg (2005) and Newman (2005).

Wheeler's supervision between the mid-1950s and the mid-1960s, and played a role in the development of the field, are Komar in 1956, Misner in 1957, Dieter Brill in 1957, John R. Klauder in 1959, Fred K. Manasse in 1961, Richard W. Lindquist in 1962, and Kip S. Thorne in 1965. Moreover, Joseph Weber spent one academic year working in close collaboration with Wheeler in 1956–1957.

Princeton University appeared not to play such a strong role as Syracuse as a center for postdoctoral education in theoretical issues related to general relativity. However, there were important developments related to in-house research on general relativity pursued by other scientists in close collaboration with the Wheeler group. Among those who worked in the field at Princeton were Martin Kruskal up until 1959, Penrose in 1959–1960, Hendrik van Dam in 1962–1963, besides, of course, Wheeler's long-term collaborators Misner, who stayed in Princeton until 1963, and Thorne who left Princeton in 1966. Together, Misner, Thorne, and Wheeler wrote one of the most famous textbooks in general relativity, published in 1973 with the title *Gravitation* (Misner et al. 1973, Kaiser 2012).

Wheeler's activity also had a strong impact in terms of promoting the field of general relativity in the larger physics community in the United States as is clear from the number of letters of recommendation for other scholars belonging to the emerging GRG community which can be found in Wheeler's archival collections.[8]

Wheeler was the first scientist to establish a research center in general relativity at Princeton University, but his group did not remain the only one at the time of the renaissance. When Wheeler was strengthening his group, Robert H. Dicke (1916–1997) began establishing one of the first research groups devoted to the empirical study of gravity physics as of 1956 with the plan to repeat the Eötvös experiment made possible by the considerable technological advances in recent years. Dicke's activity was instrumental in establishing and shaping the field of "experimental gravity physics" between the late 1950s and the late 1960s. Among his Ph.D.'s in topics related to GRG between the late 1950s and the early 1906s were Carl Brans in 1961, James Peebles in 1961, and Carrol Alley, William Hoffmann, Kenneth Turner, and James Brault, all in 1962 (Peebles 2017). The work with Carl Brans proposing an alternative, scalar-tensor theory of gravitation called Brans-Dicke theory would have a particularly strong impact on the community as it provided further motivation to design and perform tests on gravitational theories.[9] While Wheeler's role in the community-building activity was already recognized in the 1950s, as he became one of the founding member of the ICGRG, Dicke's role remained more marginal and he did not become a member of the ICGRG, which indicates that theoretical work was considered to be more important in the establishment of the community in its earlier stages.

[8]See, for example, Wheeler to Aage Bohr, 24 April 1956, JWP, Box 15; Wheeler to Shearin, 29 November 1962, JWP, Box 7; and Wheeler to the European Office of Aerospace Research, 5 February 1962, JWP, Box 18.

[9]On the history of this and similar theories, see Goenner (2012).

A.14.3 Stevens Institute of Technology, Hoboken

James L. (Jim) Anderson, one of the earlier Ph.D. students of Bergmann's, was appointed an assistant professor at the Stevens Institute of Technology in 1952. He soon established a research agenda on general relativity and gravitation theory, which gained momentum when Ralph Schiller joined the institution in 1954. It was a smaller research center compared to the major ones of Syracuse and Princeton. Nonetheless, this small institution played a major role in building the community and in redefining the research programs on GRG in the United States with the organization of the "Stevens Relativity Meetings" from 1958 onward. These were meetings held once a month at the Stevens Institute in which the most pressing problems in the emerging field of GRG were addressed by the attendees. According to some of the protagonists interviewed, these events were of enormous importance in the socio-epistemic process of the renaissance as they were conducive to the discussions between the growing number of scholars working on the East coast of the United States (as mentioned above, Syracuse, Princeton and the Stevens Institute, plus the RIAS, the IOFP, and the University of Maryland).[10] In addition to the organization of the Stevens Relativity Meetings, this research center also produced at least one Ph.D. in the field in the period between the mid-1950s and the mid-1960s (John Stachel in 1962 under Anderson's supervision) and was for a long time the home institution of David Finkelstein (from 1953 to 1960).

A.14.4 Purdue University, Lafayette

Because of its early appearance in the American pre-renaissance period, this center should be mentioned, although it was essentially limited to the activities of Dutch physicist Frederik J. Belinfante (1913–1991). From 1952, Belinfante began pursuing a research agenda aimed at the quantization of the Einstein equations of gravitation. As emphasized by historians of science Alexander Blum and Thiago Hartz, Belinfante pursued different approaches, including the canonical quantization of Einstein's theory, the quantization of the linearized theory, and the quantization of the interaction between the gravitational field and other fields (Blum and Hartz 2017). Notwithstanding its early appearance, this center remained quite isolated, both geographically and intellectually, and did not become a central institutional player in the GRG community. It was only joined, for three years, by

[10]Dean Rickles and Donald Salisbury, interview with Louis Witten, 17 March 2011, https://www.aip.org/history-programs/niels-bohr-library/oral-histories/36985. Accessed 12 March 2017. See also Dean Rickles and Donald Salisbury, oral interview with Jim Anderson, 19 March 2011; and Dean Rickles and Donald Salisbury, oral interview with Dieter Brill and Charles Misner, 16 March 2011. I am very grateful to Dean Rickles and Don Salisbury for giving me access to the recorded interviews.

Gupta from 1953 to 1956. One indication of this isolation is that Belinfante did not become a member of the ICGRG when it was established.

A.14.5 Institute of Field Physics (IOFP), University of North Carolina at Chapel Hill

The Institute of Field Physics, established in 1955 under the joint directorship of Bryce DeWitt (1923–2004) and his wife Cécile DeWitt-Morette (1922–2017), was the first university research institute specifically dedicated to the field of gravitation theory in the United States. The venture was financially supported by a wealthy industrialist, Agnew Bahnson, who was intrigued by gravitational physics and fascinated by the possible outcomes of research in gravitational theory for the development of anti-gravitational devices. The major research agenda of the IOFP was in the area of covariant quantization schemes of Einstein's equation—an approach Bryce DeWitt had been developing since his Ph.D. thesis under Julian Schwinger's supervision at Harvard in 1950 (DeWitt-Morette and Rickles 2011). The GR1 conference "On the Role of Gravitation in Physics" held in January 1957 was organized as the inaugural event for launching this new institute (see Sect. 4.2). The event was of major importance for the development of the field in the United States and beyond as well as for the formation of the community.

In the early period, DeWitt had only one graduate student working in the field of GRG, Robert W. Brehme (from 1956 to 1959). Like Syracuse, the IOFP was very successful in attracting young postdocs and research associates from different countries and providing a favorable environment for this type of research. In the period between the mid-1950s and the mid-1960s, the IOFP hosted, among others, Bertel E. Laurent in 1957–1958, Felix Pirani in 1958–1959, Robinson in 1959–1960, Leopold E. Halpern in 1960–1961, Frank R. Tangherlini in 1960–1961, Ryoyu Utiyama in 1960–1961, Frigyes Károlyházy in 1963–1965, Giorgio Papini in 1964–1966, and Peter Higgs in 1964–1966. The Dutch theoretical physicist Hendrik Van Dam, who came to the IOFP as a research associate in 1960, later obtained a permanent position at the University of North Carolina.

Although younger than Bergmann and Wheeler, in the late 1950s, Bryce DeWitt had already established himself as a central player within the community of GRG experts, particularly in view of the success of the Chapel Hill conference. DeWitt was the third American founding member of the ICGRG in addition to Bergmann and Wheeler.

A.14.6 Research Institute for Advanced Studies (RIAS), Baltimore

In 1955, the aircraft and aerospace manufacturing firm Glenn L. Martin Company established a research institute to support fundamental research with the explicit aim of developing anti-gravitational devices. Given the general ambitious goal of the venture, it was natural to make general relativity the principal field of study of the new research center. Head of the group was Louis Witten (b. 1921) who, as he stated in his recollections, did not have any real background in gravitational theory when he accepted the position.[11] Once established, however, the RIAS became an attractive center for postdocs. Among the international researchers visiting the RIAS in the period between the mid-1950s and the mid-1960s were Dennis W. Sciama in 1958 and Pirani in 1958–1959. The area of research pursued at this center was less clear-cut than that pursued at other American centers at the time. Witten, in particular, was free to explore alternative methodologies, such as the spinor approach to general relativity, thus becoming a precursor of one of the most successful theoretical achievements of the period, soon to be developed by Penrose. A major achievement was the publication of the edited volume *Gravitation* in 1962 (Witten 1962), whose fresh look at a variety of topics helped redefine the field of general relativity and gravitation as a physics discipline (see Sect. 5.2).

A.14.7 Aeronautical Research Laboratory, Wright-Patterson Laboratory, U.S. Air Force, OH

In 1956, one year after the Bern conference, the Wright-Patterson Laboratory of the U.S. Air Force in Ohio established a research group on gravitation headed by Joshua Goldberg (b. 1925), one of the earliest American Ph.D.'s in physics on gravitation theory in the post-WWII period. The main task of this group was to provide financial and logistic support to research centers working on general relativity and related fields (see Sect. 4.2). Besides this activity, Goldberg's section also established an in-house research center in which young physicists had the opportunity to spend a long-term period of research. Those included Newman from 1958 to 1959, Sachs from 1960 to 1961, Kerr from 1960 to 1962, and Havas from 1961 to 1962.

A.14.8 Other American Research centers

The abovementioned centers were those that played a greater or smaller role in the renaissance process and were perceived as the research centers where an active program in areas linked to general relativity and gravitation was pursued. They were not the only ones, however. Other isolated senior scholars, particularly mathematicians, continued to pursue research in areas such as mathematical questions of general relativity and unified field theory, but their research remained

[11] Dean Rickles and Donald Salisbury, interview with Louis Witten, 17 March 2011, https://www.aip.org/history-programs/niels-bohr-library/oral-histories/36985. Accessed 12 March 2017.

quite isolated and relatively unaffected by the renaissance process. To mention only one of most relevant players, the Czech-American mathematician Václav Hlatavý (1894–1969) joined **Indiana University** in 1948 and produced important findings on the unified field theory program in the early 1950s. However, he did not establish a research center working in the field in the renaissance period. A different story can be told for the German-American physicist Alfred Schild (1921–1977). After having earned a Ph.D. with Infeld at the University of Toronto in 1946, Schild accepted a professorship at the **Carnegie Institute of Technology**. Although he did some early work there on the quantization of the Einstein equation with his Ph.D. student Pirani between 1949 and 1951, Schild did not set up a research center. He was only able to establish a real research center on GRG after he moved to the University of Texas in 1957 (see Sect. 4.4).

A.15 USSR

The landscape of scientific institutions and the way they operated in the Soviet Union was quite different from the network of private research universities heavily supported by military funding bodies in the United States. In the Soviet Union, the directors of big institutions had enormous power to decide how to allocate funds and the Soviet Academy of Sciences had the greatest control of resources (Hall 2003). The centralized mode of organization of the Soviet political system shaped the way scientific research activities were structured.

In the field of general relativity, decisions of a philosophical and political nature also had a particularly profound effect on its unhealthy status in the early 1950s. The focus on useful science promoted by Soviet political circles made general relativity even more marginal during the low-water-mark phase than it was in other countries. Furthermore, general relativity, as well as other branches of modern physics, had been under attack for philosophical reasons, as it did not appear to fulfill the requirements of the official ideology of the party, namely, dialectical materialism. The development of the atomic bomb after World War II and the related relevance of physics, and of physicists, to matters of national security radically changed this situation. However, in the mid-1950s, there were no research centers active in the field in the Soviet Union.

Vladimir A. Fock (1898–1974) was the only major figure in physics keeping this research tradition alive, albeit with a very original interpretation of Einstein's theory of gravitation, rejecting its generally covariant character.[12] Fock was a highly respected internationally renowned theoretical physicist and had been elected member of the Soviet Academy of Sciences for his merits back in 1939. At **Leningrad State University**, he had since long established a school on theoretical physics specialized in quantum mechanics. Fock began working on general

[12]Jean-Philippe Martinez, Ph.D. dissertation on Vladimir Fock prepared at the University Paris 7 - Paris Diderot, to be defended in 2017.

relativistic problems in the late 1930s and focused more closely on this area in the 1950s, pursuing the attempt to develop his own idiosyncratic approach, which he believed to be consistent with Marxist philosophy. In this period, however, he did not create a research center but worked mostly alone on his non-covariant theory of gravitation. This is not to say that Soviet physicists were not acquainted with the theory. As the second book in their textbook series *Course of Theoretical Physics*, Lev Landau and Evgeny Lifshitz had in fact published in 1941 under the title *Classical Theory of Fields* what was unanimously perceived one of the best presentations of relativity theories for physicists. When translated into English in 1951, the book was extremely successful (Landau and Lifshitz 1951). Nevertheless, by the mid-1950s, apart from Fock's research, the field was not an active one in the USSR.

In the field of pure mathematics, the situation was slightly better. It is, indeed, possible to identify at least one emerging group actively engaging in problems connected to Einstein's theory. At **Kazan State University**, the mathematician Aleksei Z. Petrov (1910–1972) made a breakthrough in 1954 with his research on the classification of Einstein spaces (now known as the Petrov classification). Petrov became professor in 1956 and was on the verge of establishing his own research group. A department under his leadership was finally established at Kazan University in 1960: the Faculty of Relativity Theory and Gravitation (Rabounski 2008). It appears, however, that the implications of his breakthrough for the interpretation of Einstein's theory in physics received more recognition in countries outside the Soviet Union (Pirani 1957).

During the renaissance period, there was increased activity in this area in the Soviet Union. The third major actor was another respected senior theoretical physicist, Dmitri Ivanenko (1904–1994), who had carried out important work in nuclear physics from the late 1920s onward. Ivanenko, a professor at **Moscow State University** since 1943, had a prominent role in establishing an institutional structure aimed at supporting research on gravitational theory in the Soviet Union. Enjoying the prestige he gained through his membership of the ICGRG, Ivanenko was able to establish the Soviet Gravity Committee (SGC) in 1959. Under Ivanenko's leadership, the SGC rapidly became the major centralized institutional framework for activities in the field, along with the Soviet Academy of Sciences, in which Fock was working. In 1961, the SGC promoted the organization of the first Soviet Gravitation Conference, held at the School of Physics at Moscow State University. Chaired by Ivanenko, the conference was the first in a long tradition of national conferences. Almost 80 scientists attended this first conference, where 83 papers were presented, which indicates how fast this field of research had grown in the Soviet Union (Garbell 1963). A few years later, Ivanenko acquired even more power in the coordination of the field by becoming the organizer of the Gravitation Section of the Ministry of Education in the Soviet Union, which organized the activities of universities, probably in rivalry with the dominant Soviet Academy of

Sciences. When the ICGRG was established, both Fock and Ivanenko became members, whereas Petrov joined the committee in 1965. Apparently, the relationship between Fock and Ivanenko was tense, and they did not seem to act in a coordinated manner to promote the field of GRG in the Soviet Union.[13] One indication of this tension is that Fock was not present at the first Soviet Gravitation Conference.

Departing from this highly centralized status, physics research centers devoted to gravitation theory somewhat similar to the ones seen in other countries did not emerge in the Soviet Union until the 1960s. However, these centers were established by and around other leading figures such as Yakov B. Zel'dovich, Vladimir B. Braginsky, and Vitaly Ginzburg and had stronger links with relativistic astrophysics and experimental gravity physics (Thorne 1994).

References

Bergia, Silvio. 2005. Il contributo italiano alla relatività. *La Matematica nella Società e nella Cultura* 8a: 261–287.

Beyler, Richard H. 1996. Targeting the organism: The scientific and cultural context of Pascual Jordan's quantum biology, 1932–1947. *Isis* 87: 248–273.

Blum, Alexander. 2016. The conversion of John Wheeler. Talk Presented at the 7th International Conference of the European Society for the History of Science. Prague, 22 September 2016.

Blum, Alexander, and Dean Rickles (eds.). 2017. *Quantum gravity in the first half of the twentieth century: A sourcebook*. Berlin: Edition Open Access.

Blum, Alexander, and Thiago Hartz. 2017. The 1957 quantum gravity meeting in Copenhagen: An analysis of Bryce S. DeWitt's report. *The European Physical Journal H* 42: 107–157. doi:10.1140/epjh/e2017-80015-8.

Bonolis, Luisa, Adele La Rana, and Roberto Lalli. 2017. The renaissance of general relativity in Rome: Main actors, research programs and institutional structures. In *Proceedings of the Fourteenth Marcel Grossman Meeting on General Relativity*, ed. Massimo Bianchi, Robert T. Jantzen, and Remo Ruffini. Singapore: World Scientific (in press).

Choquet-Bruhat, Yvonne. 2014. Beginnings of the Cauchy problem. arXiv:1410.3490 [gr-qc].

DeWitt, Bryce. 1957. Principal directions of current research activity in the theory of gravitation. *Journal of Astronautics* 4: 23–28.

DeWitt, Bryce S. 2017. Exploratory research session on the quantization of the gravitational field. *The European Physical Journal H* 42:159–176. doi:10.1140/epjh/e2017-80016-0.

DeWitt-Morette, Cécile, and Dean Rickles (eds.). 2011. *The role of gravitation in physics: Report from the 1957 Chapel Hill Conference*. Berlin: Edition Open Access.

Einstein, Albert, Leopold Infeld, and Banesh Hoffmann. 1938. The gravitational equations and the problem of motion. *Annals of Mathematics* 39: 65–100.

Garbell, Maurice A. (ed.). 1963. *Theses of the First Soviet Gravitation Conference, Held in Moscow in the summer of 1961*. San Francisco: Garbell Research Foundation.

[13]For an extensive and in-depth analysis of the rivalry between Fock and Ivanenko, particularly within the activities of the ICGRG, see Jean-Philippe Martinez, Ph.D. dissertation on Vladimir Fock prepared at the University Paris 7—Paris Diderot, to be defended in 2017. There are also some comments in Ruffini (2010).

Goenner, Hubert. 2012. Some remarks on the genesis of scalar-tensor theories. *General Relativity and Gravitation* 44: 2077–2097. doi:10.1007/s10714-012-1378-8.

Goenner, Hubert. 2014. On the History of Unified Field Theories. Part II. (ca. 1930–ca. 1965). *Living Reviews in Relativity* 17: 5. doi:10.12942/lrr-2014-5.

Goenner, Hubert. 2016. General relativity and the growth of a sub-discipline "gravitation" in the German speaking physics community. arXiv:1607.03324.

Goldberg, Joshua N. 1992. US Air Force support of general relativity 1956–1972. In *Studies in the history of general relativity*, ed. Jean Eisenstaedt, and Anne J. Kox, 89–102. Boston: Birkhäuser.

Goldberg, Joshua N. 2005. Syracuse: 1949–1952. In *The universe of general relativity*, ed. Jean Eisenstaedt, and Anne J. Kox, 357–371. Boston: Birkhäuser.

Goodstein, J. 1982. The Italian mathematicians of relativity. *Centaurus* 26: 241–261.

Hall, Karl. 2003. Europe and Russia. In *The Oxford companion to the history of modern science*, ed. John Heilbron, 279–282. Oxford: Oxford University Press.

Hentschel, Klaus. 1997. *The Einstein Tower: An intertexture of dynamic construction, relativity theory, and astronomy*. Stanford: Stanford University Press.

Hoffmann, Dieter. 2017. In den Fußstapfen von Einstein: Der Physiker Achilles Papapetrou in Ost-Berlin. *Deutsch-griechische Beziehungen im ostdeutschen Staatssozialismus, 1949–1989*, ed. Konstantinous Kosmas. Berlin: Romiosini. (in print).

Hoffmann, Dieter, and Mark Walker. 2007. Der gute Nazi: Pascual Jordan und das dritte Reich. In *Pascual Jordan (1902–19080): Mainzer Symposium zum 100. Geburstag*, MPIWG Preprint 329, ed. Jürgen Ehlers, Dieter Hoffmann, and Jürgen Renn, 83–112. Berlin: Max Planck Institute for the History of Science Preprint series.

Kaiser, David. 2012. A Tale of two textbooks: Experiments in genre. *Isis* 103:126–138. doi:10.1086/664983.

Kennefick, Daniel. 2007. *Traveling at the speed of thought: Einstein and the quest for gravitational waves*. Princeton, N.J.: Princeton University Press.

Kragh, Helge. 1996. *Cosmology and controversy: The historical development of two theories of the universe*. Princeton, NJ: Princeton University Press.

Lalli, Roberto. 2016. "Dirty work", but someone has to do it: Howard P. Robertson and the refereeing practices of Physical Review in the 1930s. *Notes and Records: The Royal Society journal of the history of science* 70: 151–174. doi:10.1098/rsnr.2015.0022.

Lalli, Roberto, and Dirk Wintergrün. 2016. Building a scientific field in the Post-WWII Era: A Network Analysis of the Renaissance of General Relativity. Invited talk at the Forschungskolloquium zur Wissenschaftsgeschichte, Technische Universität, Berlin, 15 June 2016.

Landau, Lev Davidovich, and Evgeny M. Lifshitz. 1951. *The classical theory offields*. Cambridge, MA: Addison-Wesley Press.

Lichnerowicz, André. 1992. Mathematics and general relativity: A recollection. In *Studies in the history of general relativity*, ed. Jean Eisenstaedt, and Anne J. Kox, 103–108. Boston: Birkhäuser.

Misner, Charles W., Kip S. Thorne, and John Archibald Wheeler. 1973. *Gravitation*. San Francisco: WHFreeman.

Møller, Christian. 1952. *The theory of relativity*. Oxford: Clarendon Press.

Nastasi, Pietro, and Rossana Tazzioli. 2005. Toward a scientific and personal biography of Tullio Levi-Civita (1873–1941). *Historia Mathematica* 32: 203–236. doi:10.1016/j.hm.2004.03.003.

Newman, Ezra T. 2005. A biased and personal description of GR at Syracuse University, 1951–1961. In *The universe of general relativity*, ed. Jean Eisenstaedt, and Anne J. Kox, 373–383. Boston: Birkhäuser.

Peebles, Phillip James Edwin. 2017. Robert Dicke and the naissance of experimental gravity physics, 1957–1967. *The European Physical Journal* H 42: 177–259. doi:10.1140/epjh/e2016-70034-0.

Pirani, F.A.E. 1957. Invariant formulation of gravitational radiation theory. *Physical Review* 105: 1089–1099. doi:10.1103/PhysRev.105.1089.

Rabounski, Dmitri. 2008. Biography of Alexei Petrov (1910–1972). *The Abraham Zelmanov Journal* 1: 27–29.

Rosenfeld, L. 2017. On the quantization of wave fields. *The European Physical Journal H* 42: 63–94. doi:10.1140/epjh/e2016-70041-3.

Ruffini, Remo. 2010. Moments with Yakov Borisovich Zeldovich. In *The Sun, the Stars, The Universe and General Relativity: International Conference in Honor of Ya.B. Zeldovich's 95th Anniversary, AIP Conference Proceedings*. 1205: 1–10. doi:10.1063/1.3382329.

Salisbury, D. C. 2012. Peter Bergmann and the invention of constrained Hamiltonian dynamics. In *Einstein and the changing worldviews of physics*, ed. Christoph Lehner, Jürgen Renn, and Matthias Schemmel, 247–257. Boston: Birkhäuser. doi:10.1007/978-0-8176-4940-1_11.

Salisbury, D., and Sundermeyer, K. 2017. Léon Rosenfeld's general theory of constrained Hamiltonian dynamics. *The European Physical Journal H* 42:23–61. doi:10.1140/epjh/e2016-70042-7.

Stachel, John. 1992. The Cauchy problem in general relativity—The early years. In *Studies in the history of general relativity*, ed. Jean Eisenstaedt, and Anne J. Kox, 407–418. Boston: Birkhäuser.

Synge, John Lighton. 1956. *Relativity: The special theory*. Amsterdam: North-Holland.

Synge, John Lighton. 1960. *Relativity: The general theory*. Amsterdam: North-Holland.

Thorne, Kip S. 1994. *Black holes and time warps: Einstein's outrageous legacy*. New York: WWNorton.

Wheeler, J. A., and Ford, K. 1998. *Geons, black holes, and quantum foam: A life in physics*. New York: WWNorton & Company.

Witten, L. (ed.). 1962. *Gravitation: An introduction to current research*. New York: Wiley.

Appendix B

Full text of the letter from André Mercier to Relativists throughout the World, November 1972 ISGRGR, in which Mercier announces the proposal to establish the society, send the draft constitution, and ask for comments and amendments.

November 1972

To Relativists
throughout the World

Dear Colleagues,
At the International Conference GR6 on General Relativity and Gravitation held at Copenhagen in July 1971, an ad hoc Assembly of Scientists adopted a Resolution according to which

(i) The creation of an International Society on GRG was recommended,
(ii) a Special Statute Committee was elected and instructed as to prepare a Draft Constitution (Statutes) of the said Society, which has be done,
(iii) various other steps (election of 8 members of the International Committee on GRG, etc.) were taken.

The said Assembly expressed the wish that the Draft Constitution be circulated among all known interested scientists long before a new Assembly could formally approve the text, in order to allow for amendments to be taken into consideration.

The Statute Committee has instructed me as to submit the Draft to a lawyer in order to have a text which is in agreement with the legislation of the country where the seat of the Society is meant to be chosen. The necessity to have a juridically correct wording is urgent, for in case controversies should arise, and we cannot exclude the case, even among scientists, we must be in a position to apply the Statutes correctly.

It is my pleasure today to send you all a copy of the Draft Constitution as it stands now.

If a friend of yours unknown to us wants to have a copy, let him simply ask for it at the above address.

R. Lalli, *Building the General Relativity and Gravitation Community During the Cold War*, SpringerBriefs in History of Science and Technology,
DOI 10.1007/978-3-319-54654-4

You are all invited to comment upon this Draft. But please think first that if I receive a few hundred amendments, the task will be quasi impossible to revise again. The Statute Committee has taken care of the general wishes of the Assembly and considered carefully the limit conditions under which such a Society shall work. The lawyer, who is juridic Counselor of one of the foreign Embassies at Berne (Switzerland) has quite an experience of these things. It is possible that the English needs still some improvement.

Hence, be moderate, please, in your suggestions, and consider that our Society has to work for Scientists from all countries of the world.

The International Committee on GRG shall meet in June 1973 in Paris. You are welcome to send your comments until the end of April 1973 to me. Thank you.

Finally, I may recall that our colleagues in Israel will invite to the GR7 International Conference in 1974. This conference has been announced to IUPAP (International Union of Pure and Applied Physics).

We cannot yet make final announcements as to exactly how the Assembly which shall ratify the foundation of the new Society will be invited to convene. The Committee on GRG will decide upon that at its Paris meeting.

In the meantime, I send you all my best wishes and remain,

Sincerely yours,
André Mercier
Secretary to the International
Committee on General Relativity
and Gravitation

Printed in the United States
By Bookmasters